AF776443

Fabulous Ireland
Ibernia Fabulosa

FABULOUS IRELAND
IBERNIA FABULOSA

Imagining Ireland in Renaissance Italy

ERIC HAYWOOD

PETER LANG
Oxford · Bern · Berlin · Bruxelles · Frankfurt am Main · New York · Wien

Bibliographic information published by Die Deutsche Nationalbibliothek.
Die Deutsche Nationalbibliothek lists this publication in the Deutsche Nationalbibliografie; detailed bibliographic data is available on the Internet at http://dnb.d-nb.de.

A catalogue record for this book is available from the British Library.

Library of Congress Control Number: 2013956716

Cover illustration: Nicolaus Germanus after Claudius Ptolemaeus, *Cosmographia Claudii Ptolomaei Alexandrini*, c.1467, National Library of Poland.

ISBN 978-3-0343-1758-0

Hochfeldstrasse 32, CH-3012 Bern, Switzerland
info@peterlang.com, www.peterlang.com, www.peterlang.net

This publication has been peer reviewed

'Ciò che piace al più possente è giusto'
('Whatever pleases the most powerful is right')
—Torquato Tasso, *Re Torrismondo*, Act 5, Scene 1

Contents

Apologies and Acknowledgements

As the Introduction explains, this book is about language-games, but it is also a language-game itself, as are all our spoken and written communications. The kind of game it probably resembles most is an old jigsaw puzzle, which for a very long time has lain forgotten in the attic and from which, as a result, many pieces are still missing. I hope to have assembled the pieces I was able to recover in such a way that the puzzle makes sense to my readers. However I am fully aware that the pieces could well have been assembled otherwise, with the puzzle taking on a different shape. Indeed, it is my hope that at least some of my readers will attempt to do just that, for that is how language-games are to be played. We all construct reality differently.

Nevertheless, in order to be understood, there are rules we should all obey. Yet I must confess that there are some rules I find it very difficult to learn, or relearn. That may be due to my demographic profile. I therefore prefer to stick with BC and AD, to talk about the 'discovery' of the 'New World', and to use 'he' when referring to a person of undefined sex (for instance, the reader). But I do so for the sake of convenience, not out of malice or prejudice. I therefore pray my readers to bear with me, especially if their profile is different from mine.

I also apologize for any other shortcomings the book may have. They are possibly the fruit of my stubbornness. They are certainly not the responsibility of the friends and colleagues who were kind enough to read the book at various stages on its way to completion: Jennifer Petrie (of what used to be known as the Italian Department at University College Dublin), Corinna Lonergan FTCD (of what is still known as the Italian Department at Trinity College Dublin), Brian Richardson (of the School of Modern Languages and Cultures at the University of Leeds) and my friend and former student Paul Finucane. I am extremely grateful to them for their assistance. Their reactions, comments and suggestions, as well as those of anonymous readers, have been invaluable.

I am also very grateful to John Richmond, Professor Emeritus of Greek at UCD, and Anna Chahoud, Regius Professor of Latin at TCD, for their patient and learned assistance with translations from the Latin, which was all the more appreciated for confirming my suspicions that the humanists' grasp of classical Latin could sometimes be as hesitant as mine. But though they assisted me, they are not to blame. Unless otherwise indicated, all the translations in the book, from the Latin as well as from the Italian, are my own. Last but not least, I would like to extend heartfelt thanks to Christabel Scaife of Peter Lang. A better editor is impossible to imagine.

The book was born in Greece, on my beloved island of Spetses, where I had the privilege of spending a full sabbatical year. It is therefore dedicated to the island and its inhabitants. Perhaps it will encourage some of my readers to visit that other *fabulosa* place as well.

Eric Haywood
Head of Italian Studies
University College Dublin

March 2014

The author and the publishers wish to acknowledge the generous support received towards the publication of this book from the National University of Ireland and from EMRO Finance Ireland Ltd (Banca Popolare dell'Emilia Romagna Banking Group).

INTRODUCTION

Whose World? The Need not to Know

Usually, when we say that someone or something from the distant past is modern, we mean it less as a statement of fact than as a compliment. But it is a back-handed compliment, paid out of pique. What we really mean is: is it not amazing – and annoying – that someone or something from so long ago is so much like us? The realization is unsettling. It undermines our belief in our uniqueness and in progress, on which most of us have built our view of human history. The history of mankind, or at least its historiography, should mean a history of ever greater enlightenment, since those who live today know – or can know – so much more than those who lived, say, five hundred years ago. Their understanding or their insight should therefore be so much more developed. But if that is not so and if ideas we believe are original to us turn out not to be, then are we really who and what we think we are? And who, then, are They? Are They not Different? Are They not Other?

Since that, in essence, is what this book is about – ideas about the Self, ideas about the Other, History that does not necessarily move forward, Progress that is not – there could be no better way to start than with a compliment of the sort, namely the account of a story told in the early sixteenth century that is so modern that, with great sophistication, it puts forward views regarding travel writing and encounters with or descriptions of strangers that some today would hold to be the brainchild of modern academics. It is a story told in the *Novelle* of the friar turned bishop Matteo Bandello (1485–1561), a latter-day Boccaccio whose work was to be almost as popular as the *Decameron* it sought to imitate, and it is a story which has as its protagonist the only Italian Ireland-watcher of the Renaissance actually to have visited the place. We will find out more about him in Chapter Four. His name is Francesco Chiericati (c.1480–1539). He was a bishop, originally from Vicenza in northern Italy, who in 1516 was sent as papal nuncio to the court of Henry VIII of England and who, the following year, took advantage of an outbreak of the plague in London to escape to Ireland and pay a visit to St Patrick's Purgatory, on Lough Derg in what today is Co. Donegal, as he had been instructed to do by his friend and patroness Isabella d'Este, the imperious Marchioness of Mantua.

In the story it is another imperious lady, the 'illustrious and valorous' Lodovica Sanseverino Landriano, Countess of Pandino, who makes him give in to her wishes. The story is set on her country estate, Pandino, where Chiericati, a much-travelled gentleman, had stopped off on his way from Portugal to Rome, in order to pay his respects to his friend Alessandro Bentivoglio and to the latter's wife, 'the most virtuous heroine, the Lady Ippolita Sforza', who were among the house guests of the Countess.[1] Among them too was Bandello, and it is he who tells the story. It is thus presumably a true story. But truth, as we shall find out in the course of this book, is a moving target. More often than not it is fashioned according to the needs of those whose interests it is meant to serve. 'Whatever pleases the most powerful is right', as it says above, in the epigraph. Or in other words, might is right. And what is right is true.

In Portugal – where he had been sent by the pope, in 1521, with the charge of enlisting the support of the Portuguese to convert the Ethiopians – Chiericati had collected much information about 'the voyages made every year on the orders of that king [of Portugal], to the islands that are in that new land, where day by day his empire grows happily'; and the Countess was eager to hear all about it. She therefore pressured Chiericati into deferring his return to Rome just a little longer, so that he could entertain the denizens of Pandino with his tidings of the 'new'. At first Chiericati was reluctant to agree – after all, he was on urgent business for the propagation of the faith – but so insistent was the Countess and so inviting her estate, that in the end he gave in; and in order to make sure he could not change his mind and 'would not think, that day, of leaving them', the Countess then invited him, before beginning to recount his tales, to divest himself of his riding gear and partake of mass and luncheon with the other guests.[2]

Having thus been initiated into the collective rituals of the *Pandineschi*, Chiericati starts to talk of islands where all the men are handsome and all the women beautiful, even though their flesh is olive-coloured, and where, 'be it summer or winter, both men and women always go naked'. To make the story come to life, he also showed his audience some 'golden objects, precious stones and other beautiful things brought from those lands', as well as 'idols skilfully made of mosaics, which those peoples used to adore, who now for the most part have become Christian'. But 'that which astounded us all', Bandello remarks, 'was that he told us of a very unusual and possibly unheard-of custom'. When a stranger comes to one of their villages, six or seven of the villagers, who are known to have the most beautiful wives, present them to the stranger so that he may choose the one he prefers and spend the night with her; and the husband of the one so chosen feels honoured and is held in great esteem by the other men, for 'jealousy has no place among those very simple and unpolished people, nor does it ever put weapons into their hands'. So astounded are the Countess and her friends by such news that they burst out laughing – which is what they like to do the most. But seemingly the 'unheard-of custom' also touches a raw nerve in them, for no sooner has Chiericati recounted it than one of the other guests, Master Tommaso Castellano, in an act of unseemly discourtesy for such a civil gathering, interrupts him.[3]

Castellano was afraid, it would seem, that the narration might get out of hand, or that the laughter could misfire. What do they believe would happen – he therefore enquires of his companions – if Gandino from Bergamo were to arrive in those islands with his wife? Do they think he could stay the course? 'I for my part', he ventures, 'am inclined to believe that if the emperor, let alone a simple stranger, were to arrive over there, Gandino would never present his wife to him, nor worry about being more appreciated than the others'. This too causes everyone to laugh; but now they are on safer ground. They laugh not because what they are told is 'new', but rather because it is all too familiar: 'because the strange and suspicious nature and the jealousy of the man from Bergamo was fully known' to them. The ground is all the safer in that the *Pandineschi* are able to revert to their own lexicon, in which a stranger is so by virtue not of being an outsider looking in on newness with bemusement, or being looked at with bemusement, but by virtue of his strange behaviour. Better still, they have succeeded in turning the joke on Chiericati who, man of the world though he is, has never, by the looks of it, heard of Bergamo, which lies a mere twenty miles up the road from Pandino, nor knows anything at all about the 'fully known' antics of its strangest citizen. 'Seeing the company full of laughter', he is thus reduced to silence and can do no more than sit and listen, while a story of a different 'new' is now told to him. In the Italian of the day, it should be noted, *nuovo* ['new'] also means unusual, strange, bizarre.[4]

The story has a very simple plot. It tells of how Gandino, 'Mr Melon without taste', was so jealous of his pretty wife, Zanina, that he tried as hard as possible never to lose sight of her or let her be in the company of other men, but how in the end Zanina, who 'spent all day long in her room with Petrarch, the *Decameron* or the [*Orlando*] *Furioso* [by Ariosto], which had just been published', still managed to dispatch him 'to the Kingdom of Cornwall' – to that metaphorical place, in other words, whose very name evokes the horns [*corna*] that grow on the forehead of a cuckold – although she did so, it must be said, with the assistance of the most unlikely of paramours: a big fat groom turned lute master who had smelly feet and bad breath. The real purpose of the story, however, is not so much to make fun of Gandino as it is to demonize the inhabitants of Bergamo. Gandino's real failing is less that he is a jealous husband than that he is a *Bergamasco*.[5]

The *Bergamaschi* represent everything the *Pandineschi* loathe and fear. Today Castellano's tirade against them would unleash upon his head the righteous indignation of the guardians of political correctness. The *Bergamaschi*, he says,

> are for the most part overly suspicious, envious, contrary, riotous, badmouthed, snoops and gossips, and always full of strange ideas, with a thousand other faults and failings, of which just one would ruin any man, even though he were full of every other goodness; so much so that if two men from Bergamo were together at a court, they might easily cause conflagration and confusion and quite turn it upside down with their extravagances, their fanciful notions and their imaginary inventions.[6]

It is in the name of his own and his friends' idea of political correctness, however, that Castellano dares to be so forthright. In their view what was most correct and guaranteed like nothing else to 'raise earth to heaven', as Giovanni Botero, one of the major late sixteenth-century cosmographers (whom we shall meet in Chapter Six), was to say, was life at court – a life which, even as Francesco Chiericati mingled with the *Pandineschi*, the famous writer Baldassare Castiglione (1478–1529) was setting out to eulogize in what was to become his enormously successful and iconic *Book of the Courtier* [*Libro del cortigiano*]. To live the perfect life of courtiers is what the *Pandineschi* strove to do, for in their mind, as indeed in the mind of the ruling class of the time, it was the only life worth living. The *Bergamaschi*, on the other hand, 'rarely choose to become courtiers, for they do not really have what it takes to carry out courtly duties'. Instead they prefer 'to weal and deal'. Worse still, where life at court is built upon the expectation of strict conformity (preferably imposed by an imperious lady, as happens in *The Book of the Courtier*), the *Bergamaschi* insist on having ideas of their own. They have 'imagination' and 'invention', which could only rhyme with 'conflagration' and 'confusion'. Let a *Bergamasco* loose, and the whole world (of the court) would be 'turned upside down'.[7]

Yet though the *Pandineschi* loathed and feared what the *Bergamaschi* stood for, they also needed it, more so than anything else. The *Bergamaschi* were the reassuring Other which confirmed the *Pandineschi* in their opinion of themselves. They were the guarantee that there could be no other place worth looking for beyond the walls of courts and estates, and that curiosity could not but spell deception, only ever leading one to the Kingdom of Cornwall. It is for that reason, above all, that the *Pandineschi* cannot bear to listen to tales about those 'islands that are in that new land'. Such tales, and such islands, and such newness, threaten to expose them to a dangerously subversive Other, which might truly spell the end of their illusions about the rights of the 'noble minds' living in the world's 'noblest part', to use Botero's words once again. That Other was all the more subversive for being like a clone of the Self. 'Jealousy', according to Chiericati, 'has no place among those very simple and unpolished people', but the lack of jealousy is precisely what the *Pandineschi* congratulate themselves about and use as a gauge by which to distinguish themselves from the *Bergamaschi*. What is more, by their own admission the *Pandineschi* engage in the very kind of wife-swapping which causes them to burst out laughing when they hear Chiericati calling it a 'very unusual and possibly unheard-of custom'.

'Now, you know very well', Castellano reminds his audience at one point,

> that when the ladies' maidens, on feast or other days, are not doing anything and an honoured guest comes to the house, so as to honour and to entertain him we dance, we play an instrument, we sing, we play amusing games and we have a good time, chatting happily together of various things; and though there is no love, it is nevertheless the custom, ordinarily, for all the gallant gentlemen to be the servants of the maidens and to serve and honour them, taking one as a sister, another as a sister-in-law, yet another as a daughter, and another as an aunt or sometimes, for a joke, as a wife, and, with roles of the sort, to entertain one another and give each other favours.[8]

For their part, the *Bergamaschi* consider such behaviour unacceptable. Why, Gandino will not even allow his wife to 'converse with strangers' on such occasions and – believe it or not – he will complain loudly to his employer that 'it is not right that [wives] become so intimate with strangers and with others'.[9]

There is much worse than that however. If Chiericati is to be believed, the inhabitants of that 'new' land 'now for the most part have become Christian' and the 'empire' of the king of Portugal, having converted them, is now growing 'happily'. But if it is possible to be happy over there – a there, significantly, which is never named: for to have a name is to exist – what then about Pandino? And if the Other is so like the Self, does that mean the Self is an illusion? No wonder Chiericati had to be shut up and be initiated into the petty *coterie* of 'illustrious, valorous and virtuous' heroines and heroes of Pandino through the rituals of disrobing, attending mass and partaking of a meal. It had all been a ploy in fact. Before going back to Rome it was necessary for him to be de-briefed, for he threatened to tell fables in a spirit which could not be coutenanced (as indeed he had been in danger of doing, we shall find out in Chapter Four, in his letter to Isabella d'Este about Ireland).

Nonetheless the spirit would not be silenced, and it is a *Pandinesco* who actually unleashes it. Twice, at the beginning and at the end, Castellano warns his fellow inmates that 'this world is like a pleasant cage full of numberless fools of different kinds, and very often those who are convinced they know the most are the least wise'. The warning could not be starker. Who can really tell – Castellano asks the other *Pandineschi* to ask themselves – what the world is really like? What is new and what is not? Who is on the outside of the cage laughing, and who on the inside being laughed at? Who tells the truth, and who the master narrative belongs to? To drive the point home, at the very end Castellano turns to Chiericati and says to him, with curious logic:

> Therefore, my most reverend lord, do not be surprised if at the name of such a notable and solemn fool this beautiful and most noble company laughed so heartily, for none of them, I believe, knows better than I do all his peccadillos and all his disgusting ways [...]. And I for my part, if I had the style, would be compelled to write an *Iliad* on the subject and send it to Rome, so that it may be consecrated to Master Pasquino or the great earl Sir Marforio. But that is to write satires and not to storytell.

In other words, the 'noble company' has laughed, but only 'I' know the truth of the matter. Yet not even 'I' can do it justice (an epic poem? a satire? a story?), and in the last resort it is Rome which will have to be the judge of it. In Rome, however, the guardians of the sentence will be Pasquino and Marforio, two ancient statues on which it was the custom to affix satirical and subversive poems (pasquinades). Put differently, it was in fact impossible to remain secluded within the confines of one's cosy world and keep the new at bay. Or was it?[10]

The striking modernity (not to say, post-modernity) of this story lies in its suggestion that there is no unique or privileged position from which to name, to narrate, to describe, not even that of the eye-witness traveller, and that there is no unique or privileged language and discourse *with* which to name, narrate or describe. There are only positions of power, and it is through the successful and strategic use of power that that privilege is constructed and becomes legitimate. In other words, the narrating eye/I must learn above all how to impose itself, or else it will be reduced to silence. It must learn to be outside the cage, laughing, so as not to be inside, being laughed at. As a result there can never be an absolute truth, which imposes itself on the basis of self-evident – observed, experienced, common-sense – propositions. Truth, as we said, is a moving target, and it is never truth *about* somebody or something, but always truth *for* somebody or something. In this case it is truth *for* the Countess of Pandino and her laughing and conforming fellow courtiers. But behind and above the Countess, as the ultimate arbiter of truth – and of the language through which it must be expressed – there stands Rome; and though the power of Rome may now be much diminished, resting as it does with Pasquino and Marforio, it is nevertheless a power which – not least because it has the last word, and is the fount of discourse – far outstrips that of the (unnamed) king whose empire 'grows happily' in the (unnamed) islands of 'that' new (and unnamed) land, where no-one has a name, everyone is naked and the difference between yours and mine is all but cancelled out. Not to have a name (not to be clothed) is not to exist. Existence is in the gift only of those who can and do name, *if* they so choose. If, that is, they choose to see, and to look at, what comes or is placed before their gaze. Existence is in the gift of empire, of the empire of Rome.

The expression 'imperial eyes' has been coined to describe this particular type of eye-sight,[11] which, as we will see, also afflicts most of the Ireland-watchers discussed in the following chapters, together with all the other side effects described by Bandello. But judging by the behaviour of the Countess of Pandino – a not untypical behaviour, as we shall soon find out – perhaps it would make more sense to call it 'imperious eyes'. In any case both words derive from the same source – *imperium*: power, authority, empire – and both, I would suggest, are equally characteristic of Renaissance humanism, which is the cultural crucible in which most of the authors who are the subject of this book were formed.

This is not the place to bring up, yet again, discussions about the Renaissance and about humanism, which have been rehearsed and re-rehearsed to the point where there is little consensus among experts any-more.[12] I take it as read that, starting in Italy, *the* Renaissance did indeed take place, between the fourteenth and the sixteenth centuries (the period covered by this book), not least because the cultural and political élites of the time – who have been called 'born-again ancients',[13] though I prefer to call them born-again Romans, because Rome meant so much more to them than Greece, especially in Italy – convinced themselves and others that they were living in a new age, which was witnessing the rebirth of the enlightenment of the ancients, following a millenium of what they called (and we then also came to call) dark or middle ages. It may all have been just rhetoric, but rhetoric, as we now know from the Countess of Pandino and her friends, is a powerful tool. It can make things happen – *and* un-happen. It has the power to 'produce the world'[14] – *and* leave it un-produced.

I also take it as read that humanism – a useful label to describe the Renaissance intellectual and artistic avant-garde (turned establishment), with their particular interest in the *human-ness* of humans, in all its forms and manifestations, but especially as brought within their purview by the learning of the ancients, handed down to them by newly (re)discovered works and monuments from ancient Rome and Greece – was inherently imperialistic. A more innocent, and possibly more common view of the phenomenon would tend to see it uniquely as a power for good, which gave a singularly positive impetus to Europe's own emerging enlightenment. It was that too of course. But the humanists of the time, of whom Bandello's Castellano is an epigone, knew only too well that ancient Rome came as a package. The rights of Rome were inextricably linked to the might of Rome. Both, as humanists saw it, were the cause and effect of one another. So humanism, which was essentially meritocratic and élitist, championed the promotion of both. The evidence I adduce in the following pages will, I hope, prove the point.[15]

If it is the case that the inherent imperialism of humanism is not highlighted as often as it possibly should be, it is also the case that there is one aspect of humanism that is hardly ever highlighted at all, in works that deal specifically with the matter, and that is its geography, which is also, tangentially, the subject of this book. The omission is surprising, for not only is Renaissance geography a field of study with a well-established pedigree, and rightly so, since the interest in geographical matters was enormous at the time, but if what characterizes the *Re*-naissance is *re*-birth and *re*-discovery (of ancient culture), there can be no doubt that little else was reborn at the time quite so radically as geography.[16] Its rebirth was epoch-making, and to it we owe, if nothing else, our modern measurements of the earth, our modern maps and the discovery of America. What is more, as we shall soon be finding out, no branch of Renaissance learning was quite as imperialistic, or better perhaps, imperializing, as geography. It may have sought to measure and map the world in objective terms, but its objectivity was nevertheless Rome- and Euro-centric. And more often than not it made no secret of the bias which upheld that objectivity. It was a geography less of what *is* than of what *matters* – to those who have *imperium*.

Humanist geography was of course, as has been rightly pointed out, a 'mixed genre'.[17] Under the broad headings of 'cosmography' (study of the whole world) and 'chorography' (study of individual parts of the world), as given currency by the rediscovered *Geography* of Claudius Ptolemy (see Chapter One), it sought to describe, to measure, to locate and to represent, and it could be mediated through a variety of forms: from maps and descriptions to poems and letters, or even mere utterances, from works of history to works of fiction, from encyclopaedias to atlases. But there is one particular form which has received more especial attention in recent years, and that is travel writing, which owes its popularity, no doubt, to its close association with discovery in an age which to many of us is known as the Age of Discovery by antonomasia, as well as to what is seen to be its seminal influence on modern, empirical science (i.e. science based on observation), and above all, I would say, because it lends itself particularly well to the playing out of our own age's post-colonial *Angst*, be it the *Angst* of the (former) colonizers or the *Angst* of the (formerly) colonized. 'Travel and colonial writing' are often mentioned in the same breath, as though they were the two sides of the same coin, and that is a view which now seems broadly to be taken for granted.[18]

Travel writing is probably the category in which it is safest to include the motley collection of texts that form the subject of this book – though not without some important provisos. Travel writing is indeed what they generally are, if by that we mean 'that varied body of writing which, whether its principal purpose is practical or fictional, takes travel as an essential condition for its production'.[19] However we should bear in mind that for the most part they tell of travels without travellers, so to speak. With one, or possibly two exceptions, they were not written by people who actually visited Ireland. Instead they depict travels of the imagination – or lack thereof, as the case may be – and that is why this book is subtitled *Imagining Ireland*, and why its title is *Fabulous Ireland, Ibernia Fabulosa*. Not only was the expression coined at the time (see Chapter Four), but the purpose of (virtually) travelling to Ireland was often just that: to put its 'fables' to the test and try and sort them out from, or possibly protect them from what Machiavelli, in Chapter Fifteen of *The Prince*, famously called 'the actual truth of things' ['la verità effettuale della cosa'].[20] Not infrequently the fiction of Ireland was preferred to the fact of it. There were times too when (not) travelling to Ireland was no more than the by-product of travels of a different sort: travels from this world to the next, from the present to the past, from the loss to the rediscovery of the Self. Or else Ireland might serve as a vehicle for discussions of island-ness: why do islands exist? what is the point of them? In short, Ireland in Renaissance Italy was the domain of armchair travellers.

Because of that, it is not possible to ascribe to Italian descriptions, or better, imaginings of Ireland in the Renaissance the same degree of significance with which other forms of travel writing have been invested. They may well be the result of curiosity, but hardly ever are they the result of that 'empirical curiosity' that is seen as typical of travel writing; and strictly speaking they are not the product of discovery, or of observation, or of encounters, or of exchanges, or of confrontation.[21] Instead what lies behind them is probably no more, or less, than what in a recent work on the humanism of the so-called Father of Humanism, Francesco Petrarca (known in English as Petrarch: see Chapter Two), has been called 'the humanist tradition [initiated by Petrarch] of care of the self'.[22] It was themselves these armchair travellers were looking out for, not really the Irish. And although the comparison may be insulting to Petrarch and his fellow humanists, they are probably best compared to today's package tourists.

When your stereotypical package tourist goes to the Costa Blanca, he does not do so in order to discover or observe the place, or to encounter its inhabitants. He goes for the sun, the sand and the sex, preferably with beer and baked beans on toast by the side. In other words, he goes to look after, to care for himself. What he may do, of course, is go with thoughts of confrontation on his mind because, like the Countess of Pandino, he may be afraid of the Other and wish to demonize it, in order not to be challenged in his belief about himself. Like the *Pandineschi*, what he fears above all – and in that he is probably not unlike most of the travel writers of the Renaissance – is to grant to the Other the same ontological status as he does to himself. To him, as to Renaissance travel writers and above all to the *Pandineschi*, it is simply 'ideologically inconceivable that there should exist an otherness of the same ontological status as the same'.[23] That would spell the end of his own identity and of his self-importance. All he wants is that the Other be 'palatable', and remain inside the cage.[24]

Paradoxically and perversely what he takes to be his first-hand experience of the Other, but is often no more than prejudice, will then feed into what in the language of Eurocrats is known as the *acquis* – the acquired, shared, accepted body of knowledge – about the place (the Costa Blanca, Spain, the Spanish, etc.) and so take on the mantle of truth ('in Spain you *always* ... *all* the Spanish ...') and become part of the common standard according to which other travellers will experience the place and confirm the validity of the *acquis*. Something similar happened in the Renaissance with regard to the fables of Ireland. When Chiericati went there, all he was really interested in were those fables, and although his report on them to Isabella d'Este was written with a strong dose of derision, or better perhaps 'desire and derision',[25] implying that he did not really believe in them (and to call them fables meant, of course, that by definition they were the opposite of true), nevertheless, as far as he was concerned, that was all there really was to Ireland. That was the truth about Ireland, about *Ibernia fabulosa*.

To compare Petrarch and other humanists to package tourists *is* of course insulting to Petrarch – and possibly to package tourists as well. A better comparison might have been with the hippies of the 1960s and '70s, who travelled to Nepal in search of themselves, and drugs. There was of course a far greater degree of intelligence and sophistication in Petrarch than in the stereotypical package or hippy tourist, as well as genuine curiosity about the Other and the New, to the investigation of which Petrarch devoted his entire life. Equally, however, Petrarch's over-riding fascination was always with himself – with his *états d'âme*, as the French would say – and deep down there was a strong narcissistic streak to his curiosity, as we will find out in Chapter Two. So 'care of the self' is most definitely a good way of describing the frame of mind in which he, and the humanists who followed him, (virtually) travelled to Ireland and wrote about it. Indeed, that is what humanism was essentially about: 'me' *qua* 'human'. To that extent it is also correct to say that, with regard to Ireland, the Italian describing eye/I was a colonizing eye/I. It sought to people Ireland with its own fantasies and delusions. But it was not colonizing in the strict sense of the word.

With the exception of Polidoro Virgili (Polydore Vergil in English: see Chapter Five), who however was an agent of the king of England, and possibly the intriguing Fitzgeralds of Florence (see Chapter Three), or the drunken innkeeper and wayfarer of Giovanni Pontano's *The Ass* (see Conclusion), Italians of the Renaissance had no designs on Ireland, no dreams of conquering it. Some, though, may well have harboured thoughts of civilizing it, or at least held the belief that it demanded to be civilized, not least because it was in Europe (or was it?) and the Irish professed to be Christian (but were they?). And Europe, as humanists were at one in agreeing, was the best part of the world, so could not really be the home of backwoodsmen – of men incorrigibly stubborn in their backwoodishness – such as the Irish. Incidentally, this is another reason why the case of Ireland may be different from that of other Others.[26] The Irish were European Others – *Bergamaschi* so to speak – not Others from 'the islands that are in that new land', and in that they possibly posed an even greater conundrum than did Chiericati's happy natives, the conundrum so tellingly highlighted by Tommaso Castellano: how could the *Bergamaschi* possibly want to remain *Bergamaschi*, and not become courtiers? Were they not Italians after all? Yet all things told, the *Pandineschi* did actually prefer it that way. It was more reassuring. And that is why, possibly, it was also preferable for *Ibernia* to remain *fabulosa*, for its fables to be its truth. Fabulousness was certainly easier to accommodate in Europe, as European, than 'riotousness' and 'suspiciousness'.

If Italian humanists were not colonizers, with regard to Ireland, they were certainly imperializers, in the sense in which the Countess of Pandino and Tommaso Castellano are. Rome was their standard, and what Rome did was best and was right. Indeed it can surely be argued that it was Italian humanism that fostered and legitimized European imperialism, by ushering in its first, innocent phase, where empire-building was seen as beneficial, inevitable and natural. Roman civilization – the *nec plus ultra* for any humanist – was the Siamese twin of Roman *imperium*. They grew together and it was Rome's empire which brought civilization to the rest of the world. So what could possibly be wrong with it, with empire-building? Or rather, what could there be that was not right about empire-building?

Of course imperialism is never innocent, but what does characterize the imperialism of Italian humanism – which is mostly theoretical or ideological, not to say rhetorical – is the fact that, initially at least, it bread no doubts, brooked no second thoughts. To question the assumption that empire was a Good Thing would have been as absurd and unthinkable as questioning the existence of God (which, incidentally, humanists as a rule did not do either, contrary to what is sometimes believed). To them it was certainly innocent. However, to sustain and maintain that innocence they were undoubtedly to engage in what have been labelled 'strategies of innocence',[27] namely patterns of speech and behaviour designed to demonstrate that it is the imperializers – the colonizers, the aggressors – who are the wronged party. As we will see in Chapter One, this too is something they learnt from their Roman forebears. Julius Caesar, in *De bello gallico*, is very deft at arguing that in their invasion of Gaul it was the Romans who were the victims. The Countess of Pandino is equally as deft. According to her self-serving logic, it is the *Bergamaschi* and Chiericati's happy natives who have set out to be an insult to the *Pandineschi*, not the other way round.

The Countess's strategy could also be seen as an example of what has been called, specifically in relation to travel writing, a 'language-game'. What is meant by language-game (a concept originally formulated by the philosopher Wittgenstein) is

> a set of rules and assumptions, often unconscious to the agents who use them, organized so as to meet the demands of a social context of communication. A 'language-game', therefore, encompasses any cultural activity based on the interaction of subjective agents who create and exchange mental representations in order to meet various purposes.[28]

In other words, a language-game is what we all engage in when we converse, exchange views, or otherwise interact with other people, but what the expression primarily, and rightly, suggests is that communication between various interlocutors is based on a series of culture-specific intellectual or ideological assumptions and context-specific codes of communication, which the bare words of any utterance, especially if they are dead, that is, written, do not properly convey, and may even conceal. Consequently, 'the cultural historian needs to reconstruct the assumptions and rules underlying any cultural production [...], identifying the context in which (let us say) a proposition makes sense within a text, or a text makes sense within a larger body of literature'.[29]

Once again, where Bandello is strikingly modern is not only in anticipating and illustrating the concept of language-game – which is quite literally what the *Pandineschi* engage in with Chiericati, and amongst themselves – but in showing that the game was not necessarily 'unconscious to the agents' involved in it. Castellano, for one, is all too aware that his fellow courtiers were playing games, and he winks at us to warn us of the fact, conscious as he is that the game, especially if it is in written form (and thus, according to humanists, anything *but* dead), is always also played with posterity: we too, as readers, are involved in it, and so, as we try to make sense of it, we must be aware of and question our own assumptions and codes. But Castellano, of course, is not alone in this. If there was one thing humanists were fully convinced of, as they 'entered into the ancient courts of ancient men' (to use an expression of Machiavelli's: see Chapter One) and 'talked with books' (to use an expression of Petrarch's: see Chapter Two), it was the rhetorical nature of truth and reality. They saw truth and reality not as something which *is*, but as something which is *constructed*, collectively. All of them therefore consciously played language-games. And in particular, as we have seen with Bandello/Castellano, they played them with the dedicatees or target audience of their works, since for humanists, as I said, truth was more *for* than *about*.

Therefore, when humanists imagined Ireland, they imagined it *for* a particular purpose, which is why it is possible for us to talk of the Italian Ireland*s* of the Renaissance (in the plural) rather than the Italian Ireland of the Renaissance (in the singular). In any case, even though the Renaissance witnessed the emergence (not least because of the influence of humanists, who trumpeted the Italians' unique destiny as heirs to the Romans) of a sense of Italianness, especially among the cultural and political élite of the time, and even though a standard Italian language was beginning to take shape (and even though I will continue to use the expressions Italy and Italian for the sake of convenience), there was no such thing really as Italy and Italians. The Italian peninsula was divided into a number of independent states, each with its own language, traditions and citizens (Venetians, Florentines, Sicilians, etc.). So it is possibly counter-intuitive to expect there to have been a consensus in what today we call Italy on the question of Ireland, as indeed of any other country. Different people at different times needed different Irelands. Humanist geography, as I said, was a geography of what mattered – or did not – to those for whom it was fashioned.

We should also bear in mind that there is never such a thing as a country as it *really* is. A brief perusal of any guide-book should easily convince us that there are always at the very least two countries of the same name. For instance, the Italy which tourists go in search of, with very clear ideas of what it is they want to find, is often quite different from the Italy in which Italians live. All of us in fact, like the Countess of Pandino, only ever know as much about another country as we want or need to know, or better perhaps, all of us imagine other countries simply as it suits us, whether or not we have ever gone to visit them. And so it was in the Renaissance. The legend of Ireland as the land of saints and scholars is a case in point. Today there is hardly an Irish person who would not immediately and instinctively say 'saints and scholars' when asked what foreigners, especially Italians, most readily identified with Ireland in the past – as opposed to, say, Guinness and green grass today, or possibly Joyce and U2. It is certainly true that monks from Ireland were a significant presence in medieval Italy (one need only think of St Columbanus and Bobbio, or of San Cataldo – St Cathal – of Taranto), but what is surprising – or perhaps it is not – is that Irish saints and scholars, whether under that denomination or another, appear to have been completely erased from the collective memory of Italians in the Renaissance. If anything, what Ireland stood for, besides fables, was St Patrick's Purgatory, although increasingly that too came to be seen as one of its fables.

For all of these reasons I have decided that there would have been no point in attempting to verify whether what Italians imagined about Ireland was true, no point in comparing what they said with what was (insofar as it is ever possible for us to ascertain what was). That would have been tantamount to double-guessing their intentions, to telling them what they *should* have said. But who can be the arbiter of what Ireland *should* be like, or should be *held* to be like? Apart from being anhistorical, it would be like attempting to build dykes on shifting sand. In straw-polls conducted amongst my undergraduates over the past couple of years about the defining symbols and myths of Irishness (in a class about the defining symbols and myths of Italianness), not a single student under the age of twenty-five mentioned Catholicism or the Church (where a good few had the Famine or the struggle against the English at the top of their list), and all were even surprised when it was suggested to them that they *should* have done. A mere ten years ago that would have been inconceivable. Even I – a godless blow-in, as they are called in Ireland – was shocked. So why attempt to double-guess Italians who lived half a century ago?

Instead my main focus has been to try and make sense of the language-games played by Renaissance Italians in which Ireland was a piece on the board, sometimes even only a very small one. That is, I have attempted 'to reconstruct the assumptions and rules underlying [the games], identifying the context in which [...] a proposition makes sense within a text, or a text makes sense within a larger body of literature'. In doing so I may at times appear to some to have upset the proper balance between text and context, reconstructing a very big context for a very small text, as with the Fitzgeralds of Florence, for instance, in Chapter Three. I believe it to be the case, however, that the longer the text, the more it tends to speak for itself. The shorter the text, on the other hand, the more its sense needs teasing out. Similarly, it is not always easy to strike the right balance when providing historical information, of the traditional kings-and-battles kind. A bit of history is a dangerous thing, and can lead to facile assumptions being made about historical determinism. A lot of history can seem a pointless burden. Generally speaking my aim has been to provide brief historical and biographical sketches that would help those readers not familiar with the period, and the many ones possibly not acquainted with some of the more obscure texts and authors I deal with, not to lose their way. That being the case, the sketch in Chapter One, dealing with Pius II, is longer than most, since it is the first. Readers who see no point in it, can simply skip over it, and others.

History, in the sense of chronology, is of course the easiest, most recognizable and most reassuring rule in the language-game the author of a book such as this plays with his readers. An author wishes to convey meaning that is valid and readers expect a text to impart meaning that is unbiased, and chronology seems to be able to deliver both. Chronological sequence neutralizes the arbitrariness of authorial choices and concatenation provides an appealing logic for cause and effect. However, I have chosen not to play by that particular rule, and for that I apologize to my readers. There are a number of reasons for my decision. First of all the evidence shows, I believe, that there is no clear development in the way images of Ireland in Renaissance Italy developed. There was not just one image which became ever sharper with time. Instead, as I said, different people imagined Ireland differently at different times, or in different circumstances. There were competing images, which could overlap but often did not, and the same image could be produced at the end of the Renaissance as had been produced at the beginning, depending on the needs of those they were produced by and for. People, we could say, came at Ireland from different angles, and it is those angles which the different chapters explore and reflect. My approach is broadly thematic.

Another reason for my shunning chronology is that I too, like Renaissance humanists, believe that reality is not something that just *is*, but something that is *constructed*, by all those involved in language-games; and as an Italianist I believe that Italian has got it right by not making a distinction between history and story. Both are *storia*. Reality may indeed be constructed differently over time, but there is no reason why it should, or at least, Time is not the main reason why it may be. More importantly, however, I wished to avoid giving the impression that, in Renaissance Italy, there ever was an *acquis* regarding Ireland, a body of knowledge which was known and available to all. If I am lucky, such an *acquis* (I apologize for insisting with this ugly word, which is pronounced more or less like 'a key', with the stress on the 'a' – what is most useful is often least aesthetically pleasing) is something which will come into existence as a result of this book. But at the time (and let us not forget that we are talking about a timespan of two centuries and more: historians can often seem to lose sight of the fact that time moves ever so erratically and treacherously) very few people had the benefit of the whole picture. Most only ever had a partial view, and were content with that. Like the *Pandineschi*, they constructed the world strictly on the basis of the need-to-know principle. Or better, of the need-*not*-to-know principle.

However, to give chronology its due – for I am fully aware that Time is a tyrant – and to help my readers play the game, at the beginning of every chapter I include a brief chronological table of all the authors and works discussed in that chapter. And it is also true to say that, broadly speaking, my chapters do progress through time, although such progression should not be mistaken for Progress. If anything, where they may possibly convey the notion of progress (if that is what it is) is with regard to the question of the describing eye/I. It is not a given, in my view, that the traveller, and particularly the armchair traveller, necessarily saw himself 'as a figure of authority' or that from the word go he applied 'critical common sense' to his observations.[30] Having 'challenged the principle of a hierarchical order in learning', Petrarch, for personal, existential and ideological reasons, embraced the notion of a doubting eye/I, as we will see in Chapter One, and he made of doubt an article of faith, which he passed on to his followers.[31] Admittedly humanists doubted with great self-assuredness. Like modern academics, they were professionals of doubt, who saw the right to profess as the logical consequence of their duty to doubt. But doubt they did, and all the more so for having to contend with the authority of the ancients, whom they set up as unbeatable sparring partners and who they felt were constantly looking over their shoulders.

Thus we move from the establishing of that authority in Chapter One, to the 'talking with books' of Chapter Two, with all the questioning and doubting (and professing, or growing imperiousness) it brought in its wake, to the casting aside of authorial *Angst* in Chapter Three, with the consequent desecration of totalizing models and world-views. In Chapter Four, that endeavour 'to establish a meaning outside the sacred' is taken a step further, as would-be pilgrims confront the truthfulness of fables and the fabulousness of truth.[32] In Chapters Five and Six we find the describing eye/I asserting itself with renewed authority, or aggressiveness, made necessary, it would seem, by Europe's emerging religious divide and its self-assumed mission to dominate the world, while in the Conclusion we find the describing eye/I resting on its laurels, as it were, confident in the knowledge that fiction is stronger than fact and fables more resilient than truth. Or put otherwise, that the Countess of Pandino's need-not-to-know hypothesis is a workable one. But what are the chapters *really* about?

The clue is given by their sub-titles. Chapter One ('Out of This World? *Imagining Ireland with the Ancients*') presents the views of Ireland which Renaissance humanists inherited from the ancient Romans and Greeks. Although the Renaissance was characterized by a 'well marked historical consciousness', it is nevertheless true to say that its born-again Romans felt their illustrious forbears were living in their midst.[33] If anything was an *acquis* at the time, it was what the ancients said. So ancient views on Ireland are tantamount to Renaissance views, and no discussion of the subject would be complete without them. Chapter Two ('Amazing World? *Imagining Ireland with (and without) Gerald of Wales*') discusses the discovery by Petrarch of what is possibly the 'most widely read authority' on Ireland, namely Gerald of Wales's *History and Topography of Ireland*,[34] and, following what to us may seem like Petrarch's uncommonly positive reading of the text, his sharing of it with friends and colleagues. Chapter Three ('The Other World. *Imagining Ireland in the Shadow of Dante*') presents the powerful and sometimes overbearing influence which Dante's *Divine Comedy* had on Renaissance writing and understanding of the world, and consequently of Ireland. Chapter Four ('Between Worlds. *Imagining Ireland in the Footsteps of St Patrick*') examines the fascination exerted on Renaissance Italians by what to them was undoubtedly Ireland's most famous site, St Patrick's Purgatory, which has left its mark on the Italian language to this very day, but which Italians appear to have visited to 'care for themselves' more than for any other reason. Chapter Five ('Divided World. *Imagining Ireland(s) at the Time of the Reformation*') demonstrates better than any other my point that Renaissance geography was more often *for* than *about*: two Italian humanists writing at the same time construct two totally different views of Ireland, one which could possibly be labelled Protestant, the other Catholic. Chapter Six ('New World. *Imagining Ireland in a New World*') links descriptions of Ireland with the reconfiguration of knowledge which took place – or did not – following the discovery of the New World and the reorganization of geographical lore which came about – or did not – to satisfy the growing demand for cosmographies and atlases. The Conclusion ('Mad World. *Forever Imagining Ireland*') reaffirms the enduring appeal of Ireland's fables, in accordance with the Countess of Pandino's need-not-to-know principle.

And now there remains just one question to address, but possibly the most important. Why Ireland? I was asked that many a time in the course of my research, usually by Italians, who felt the topic was of marginal significance in the context of Renaissance studies, or by colleagues 'from across the water', to use a common Irish euphemism, who tended to raise an eyebrow at the thought. The simplest answer was always: why not? But hopefully the chapters that follow, and the *storie* they tell, will provide a better one.

Notes

1. 'L'illustre e valorosa signora la signora Lodovica Sanseverina e Landriana', 'la vertuosisssima eroina la signora Ippolita Sforza' (BANDELLO (1928), p. 19)
2. '... levatisi i panni da cavalcare che facesse pensiero per quel giorno non partirsi [...] Onde egli cominciò a narrare le navigazioni ch'ogni anno quel re fa fare a le isole che sono in quel paese nuovo, ove tutto il dí il suo imperio va felicemente accrescendo' (BANDELLO (1928), p. 19). Unless specified otherwise, all cross-references in the footnotes are to works listed in the bibliography, Part 1 (when the names are spelt in SMALL CAPITALS) or Part 2 (when they are spelt in lower case). The dates of historical figures, insofar as they are known, are given (when a figure is mentioned for the first time) as follows: for popes, dates of reign; for rulers, date of birth, of beginning of reign and of death; for all other figures, date of birth and death.
3. '... ramenti d'oro, perle, pietre preziose ed altre belle cose da quei paesi recate [...] alcuni idoli maestrevolmente lavorati di mosaico, che quei popoli adoravano, che omai per la piú parte son fatti cristiani [...] venne a dire d'alcune genti le quali la state e il verno vanno sempre ignude, cosí uomini come femine [...] Ma quello che tutti ci fece meravigliare [...] fu che ci narrò un costume molto nuovo e forse piú non udito [...] gelosia tra quelle semplicissime e rozze genti non ha luogo né mai mette lor l'arme in mano' (BANDELLO (1928), pp. 19–20).
4. 'Io mi fo a credere che se colá arrivasse l'imperadore, non che un semplice forestiero, che mai egli non gli presenteria la sua moglie, né si cureria esser in questo piú apprezzato che gli altri [...] tutta la brigata cominciò a ridere, perciò che da tutti la strana e sospettosa natura e la gelosia del bergamasco era pienamente

conosciuta. Monsignor vescovo, veggendo la compagnia piena di riso …' (Bandello (1928), p. 20).

5. 'Ser […] mellone senza sapore' – 'se ne stava tutto 'l dí in camera con il Petrarca, le *Centonovelle* o il *Furioso*, che di nuovo era uscito fuori' – 'nel reame di Cornovaglia' (Bandello (1928), pp. 42, 32, 44).
6. 'Sono poi per il piú troppo sospettosi, invidi, ritrosi, commettitori di risse e discordie, rapportatori, maldicenti e pieni sempre di nuove chimere, con mill'altri difetti e mancamenti dei quali un solo guasterebbe ogni uomo, quantunque pieno d'ogn'altra bontá; di maniera che se dui bergamaschi si trovassero di brigata in una corte, sarebbero facilmente atti a porla in combustione e garbuglio e voltarla tutta sossopra con i loro ghiribizzi, fantastiche chimere ed imaginarie invenzioni' (Bandello (1928), p. 23).
7. 'Essi di rado si fanno cortegiani, non essendo molto atti agli uffici de la corte' – 'molto trafficatori' (Bandello (1928), pp. 22–3). Castiglione's work portrays a series of conversations on the subject of the courtier that take place on consecutive evenings at the ducal palace of Urbino, under the supervision of the Duchess, Elisabetta Gonzaga. The literature on Castiglione is vast. For a useful, though somewhat dated introduction see Woodhouse (1978).
8. 'Ora voi sapete bene esser comune usanza che quando le damigelle de le signore, le feste e gli altri dí, sono scioperate e che vien a casa loro onorato forestiero, che per onorarlo e festeggiarlo si danza, si suona, si canta, si gioca a' giuochi festevoli e si sta sui piaceri, cicalando insieme allegramente di varie cose; ed ancora che non ci sia amore, si costuma perciò, per l'ordinario, da tutti i galanti gentiluomini far il servidore con le damigelle e servirle ed onorarle, pigliandone una per sorella, l'altra per cognata, l'altra per figliuola, l'altra per zia e talora in burla per consorte, e con simili titoli intertenersi e donarsi dei favori' (Bandello (1928), p. 30).
9. 'Non voleva Gandino che sua moglie […] con stranieri tenesse ragionamenti […] non stava bene che tanto con gli stranieri ed altri si [dimesticasse]' (Bandello (1928), p. 30).
10. 'Questo mondo è una piacevol gabbia piena d'infiniti di varia specie pazzi, e […] assai spesso coloro che piú saper si persuadeno sono i men savi' – 'Sí che, monsignor mio molto riverendo, non vi meravigliate se al nome di questo cosí notabile e solenne pazzo, e per aggiunta fieramente ingelosito, questa bella e nobilissima compagnia sí saporitamente rise, non ci essendo nessun di loro, credo io, che meglio di me non conosca tutte le sue taccherelle e tutti i suoi fecciosi modi […] Ché io per me sarei, se stile avessi, sforzato a farvi suso una *Iliade* e mandarla a Roma ché fosse consacrata a messer Pasquino o al gran barone ser Marforio. Ma questo è far satire e non novellare' (Bandello (1928), pp. 22, 46). The truncated bust of Pasquino stands in Rome's Piazza Pasquino, just off

Piazza Navona. The colossal statue of Marforio (aka Oceanus) is now located in the courtyard of Rome's Capitoline Museum.

11. See Pratt (1992).
12. Mazzocco (2006) gives a good overview of the competing interpretations. See also: Baron (1966), Garin (1965), Hankins (2000), Kraye (1996). For a more general introduction to the Italian Renaissance see Brown (1999), Hay and Law (1989), Najemy (2004).
13. Kirkham and Maggi (2009), p. 21. The expression is Kirkham's.
14. Pratt (1992), p. 5.
15. The view that 'empire appealed' to humanists is put forward i.a. by Brown (1999), p. 5. See also Brown (2000), Hörnqvist (2004).
16. The standard work on Renaissance geography is Broc (1996). See also: Aujac (1975), Bouloux (2002), Defilippis (2009), Headley (2000), Humboldt (1992), Lestringant (1991), as well as the works listed in notes 15 and 21 of Chapter One below. The works listed in note 12 above, as also Shirley and Hoeniger (1985), are a case in point regarding the omission to even consider Renaissance geography.
17. Rubiés (2007): 'New Worlds and Renaissance Ethnology', p. 172.
18. Hadfield (2004), p. 16, as one example.
19. Rubiés (2007): 'Travel Writing as a Genre: Facts, Fictions and the Invention of a Scientific Discourse in Early Modern Europe', p. 6.
20. See Conclusion, note 7.
21. Rubiés (2007): 'Travel Writing and Ethnography', p. 14.
22. Zak (2010), p. 162. On Petrarch as the Father of Humanism see for instance Mazzocco (2006), pp. 215–42.
23. Godzich, in Certeau (1986), p. xiii.
24. Bhabha (2007), p. 112.
25. Bhabha (2007), p. 96.
26. However, compare Cardini (1987) and De Anna (1994).
27. Pratt (1992), p. 7.
28. Rubiés (2000), p. xiv.
29. Rubiés (2007): 'Travel Writing as a Genre', pp. 6–7.
30. Rubiés (2000), p. xviii; Rubiés (2007): 'Travel Writing as a Genre', p. 20.
31. Fubini (2006), p. 133.
32. Rubiés (2000), p. 44.
33. E. Garin, *Italian Humanism*, quoted in Mazzocco (2006), p. 11.
34. Hadfield (1997), p. 93.

CHAPTER ONE

Out of This World? Imagining Ireland with the Ancients

Authors and works discussed in this chapter (in chronological order):[1]

- CICERO (106–43 BC), *The Orator* [*De oratore*]
- JULIUS CAESAR (102?–44 BC), *The Gallic War* [*De bello gallico*]
- VIRGIL (70–19 BC), *Georgics* [*Georgicon*] I
- STRABO (63? BC–24? AD), *Geography*
- PLINY THE ELDER (first century AD), *Natural History* [*Naturalis historia*]
- POMPONIUS MELA (first century AD), *Cosmography* [*Cosmographia*], aka *On the State of the World* [*De situ orbis*]
- TACITUS (56?–118 AD), *Agricola* [*De vita Agricolae*]; *Germania* [*De origine et situ Germanorum*]
- PTOLEMY (90?–168 AD), *Geography*
- SOLINUS (third century AD), *Multihistory* [*Polyhistor*], aka *Miscellany of Memorable Things* [*Collectanea rerum memorabilium*], aka *On the State and Memorable Things of the World* [*De situ et memorabilibus orbis*]
- ISIDORE OF SEVILLE (570?–636), *The Book of Etymologies or Origins* [*Etymologiarum sive originum libri*], aka *On the Nature of Things* [*De rerum natura*]
- ST BERNARD (1090–1153), *The Life of Malachy* [*Vita Sancti Malachiae*]
- GERALD OF WALES, aka GIRALDUS CAMBRENSIS, aka GERALD DE BARRI (1146?–1223), *The Topography of Ireland* [*Topographia hibernica*], aka *On the Marvels of Ireland* [*De mirabilibus Hybernie*] – also see Chapter Two
- ANDREA DA BARBERINO, aka ANDREA MANGIABOTTI (c.1371–1431), *The Royals of France* [*Reali di Francia*] – also see Chapter Four

- Aeneas Silvius Piccolomini (1405–64), Pope Pius II (1458–64), *On Europe* [*De Europa*]; *On Asia* [*De Asia*]; *Commentaries* [*Commentarii*]
- Niccolò Machiavelli (1469–1527), letter to Francesco Vettori – also see Conclusion

In 1458, shortly before ascending the throne of St Peter as Pope Pius II, Aeneas Silvius Piccolomini (1405–64) put the finishing touches to an innovative work *On Europe* [*De Europa*]. Focusing entirely on contemporary events and intermingling the exposition of *what* happened with the description of *where* it happened, the work was conceived as a journey which took its readers on a tour of Europe, starting in Hungary and proceeding through Greece, the Balkans, central Europe, northern Europe, Germany, France, the British Isles and the Iberian peninsula, before coming to a close in Italy. The idea of Europe it sought to put forward was also novel. More than just one of the world's three known parts, Europe, in Aeneas's estimation, was much closer to what many believe, or would like to believe, it is today: an identity. An identity shared by peoples living together in a common space, with a common destiny, and bearing a common, Christian and Roman, heritage. Not every people, though, was allowed to share in that identity. The opening sentence of the work makes that quite clear. 'What was done', Aeneas begins, 'under Frederick, third emperor of that name, by Europeans *and* island-dwellers considered to be Christian, I will here record, for the sake of future generations, as briefly as possible and insofar as it is worth remembering and known to me'.[2] The *and* says it all: islands only belong to Europe on sufferance. To be allowed to count as European, not only must their inhabitants perform deeds which are worth remembering, but they must above all be considered to be Christian.

That – today's Hibernophiles must feel – surely boded well for Ireland. Would a pope not have known that in the early Middle Ages Ireland's saints and scholars played a leading role in keeping Europe, and above all Italy, firmly in the Christian fold, following the barbarian invasions that had brought about the fall of the Roman empire? Would he not at least have heard about San Colombano (St Columbanus or Colm), who founded a renowned monastery at Bobbio, in the foothills of the Alps, to the north east of Genoa, and died there in 615? Or about his near contemporary San

Cataldo (St Cataldus or Cathal), the patron saint of Taranto, in the arch of Italy's foot, and probably the most revered Irishman in Italy? And what of San Donato (St Donatus), who was bishop of Fiesole, above Florence, from 826 to 877? Or of Dungal, who in the early ninth century taught at Pavia, the then capital of Lombard Italy, or of Petrus Hibernicus (Peter the Irishman), who was invited by Emperor Frederick II (1194–1250) to be the first rector of the university of Naples?[3] And surely he must have been aware that St Patrick's Purgatory, on Lough Derg in present-day County Donegal, was a place of Christian pilgrimage so famous in Europe, and especially in Italy, that its name – *pozzo di san Patrizio* [St Patrick's well] – was becoming a common metaphor in the Italian language (as, incidentally, it has remained to this very day) to signify a bottomless pit?[4]

But assuming Aeneas did not know about all of that, he cannot not have been aware – can he? – of the tales of adventures told by the famous Florentine storyteller Andrea da Barberino (c.1371–1431), in which Irish knights help the Holy Roman Emperor and his paladins to defend the Christian world against its Muslim enemies. Andrea's *The Royals of France* [*Reali di Francia*] was enormously popular at the time and with the invention of printing in the latter part of the century was set to become an Italian-wide best-seller, so there cannot have been many who were not aware that, in the words of Andrea,

> there were at that time few Christians in the world, at least in Europe, because only of late had Constantine and the City of Rome been baptized. And shortly before, lords and princes of the Christians had been in Britain and in Ireland [*Irlanda*], because King Uther Pendragon and King Arthur had been there, with a great and beautiful company; [...] and those of England were themselves converted to the Christian faith, so that in all the parts of the West, England and Ireland had become Christian. [...] All the other provinces were Saracen and pagan: Spain, France, Burgundy, *Gesina*. In Asia there had begun to be many Christians towards India and towards Armenia, and already in the East were there beginning to be. And because Constantine had been baptized, he had many enemies; and all the other faiths began to hate him.[5]

In Andrea's view of the world, in other words, not only were the Irish full square in (to-be-Christian) Europe, but their deeds were eminently worth remembering. The Irish were amongst the first and truest defenders of the faith.

Not everyone, however, would have been of that opinion. There was another school of thought, which saw the Irish as quite possibly the worst of Christians. Its chief exponent was Giraldus Cambrensis (1146?–1223), the Norman monk from Wales who accompanied England's King Henry II (1133–54–89) and the latter's son Lord John (1167–1216) on their expeditions to conquer Ireland and who was the chief architect of the island's ideological subjugation, by means of his much read *Topography of Ireland* [*Topographia hibernica*], a work well known to Italian humanists, as we will see in the next chapter. The Irish, Gerald wrote, were 'a barbarous people, literally barbarous':

> of all the peoples it is the least instructed in the rudiments of the Faith. They do not yet pay tithes or first fruits or contract marriages. They do not avoid incest. They do not attend God's church with due reverence. Moreover, and this is surely a detestable thing, and contrary not only to the Faith but to any feeling of honour – men in many places in Ireland, I shall not say marry, but rather debauch, the wives of their dead brothers.[6]

It was not just supporters of the Anglo-Normans, though, who took a dim view of the faith of the Irish. More than half a century before Henry II's conquest of Ireland, St Bernard (1090–1153) had written in his *Life of Malachy* that when St Malachy had been elevated to the bishopric of Connor, in 1124, he 'realized that he had been sent not to men, but to beasts'; his flock were 'Christians in name, but pagans in deed', 'they gave neither tenths nor first fruits, they did not enter into legitimate wedlock, they did not make confession. Seekers or givers of penitence could hardly be found; ministers of the altar were very few in number'.[7]

That unflattering view of the Irish continued to have its advocates in Church and papal circles, and perhaps it was to them that Aeneas had spoken before deciding whether or not to include Ireland in his Christian Europe. What is for sure is that Ireland did not make much of an impression on him. Having taken his readers (at some speed, it must be said) through England and Scotland, he then announces:

> We ought now to acquit ourselves of Ireland [*Hibernia*], which is separated from Britain by a short stretch of sea, and is partly free, benefitting from the friendship and the alliance of the Scots, and partly subject to English rule [*imperium*], but since

> nothing worth remembering has taken place there, we understand, during the period we write about, we hurry on to matters Spanish.[8]

True, Aeneas does acknowledge that Ireland may be worth a visit, or at least that his readers may have expected it to be so – 'we *ought* to *acquit ourselves* of Hibernia', he writes, as though under some obligation – but his sense of relief at being able to hurry on from it is palpable.

A similar sense of relief emanates from the pages of Aeneas's autobiography, titled *Commentaries*, where he writes about a trip to Scotland he undertook in 1435. His employer, a cardinal, had sent him on a mission to the Scottish king, and as a result he had been obliged to spend 'seven months out of this world [*extra orbem*] in the kingdom of Scotland'. He had found the place to be unbearable and totally devoid of anything he expected or would have liked to find there. It was a 'cold land', where the days in winter last no more than four hours and the cities have no walls; where there is no wine, and bread is a luxury; where kissing a woman is of less account than shaking hands with her in Italy; where the people, who are 'poor and uncivilized', like nothing better than to hear slanders on the English; where the horses, though they are good trotters, are neither groomed with iron, nor combed with wood, nor tamed with a bit; where there are sulphurous stones beneath the ground which they extract to make fires, but where, contrary to what 'Aeneas was wont to say' (unless he used the royal we, Aeneas preferred, like Julius Caesar, to talk about himself in the third person singular), there are no trees whose fruit, if it falls on the ground, starts to rot, but if it falls in the water is changed into birds, 'for when he got there and eagerly investigated the miracle, he discovered that it was a lie; or if it was true, that it had been relegated even further, to the Orkneys'.

Not surprisingly, Aeneas's return from that land of missing things, not to say his exodus, is recounted like a tale of deliverance, and his re-emergence into 'our' world like a vision and a rebirth. The rebirth took place in Newcastle, which 'they say is the work of Caesar', and

> there for the first time Aeneas felt he could see the shape of the world again and the habitable face of the earth. For Scotland, and that part of England which is close to the Scots, has nothing in common with where we live: it is wild, uncivilized and untouched by the winter sun.[9]

Put otherwise, Aeneas viewed the world through the lens of Rome, above all of ancient Rome and *imperium*-builders like Julius Caesar, and even though he was experiencing it for himself, he had no difficulty leaving his 'empirical curiosity'[10] on the back burner, so to speak – assuming he even thought in those terms – in order to put on what we might call his Rome-tinted spectacles, thereby to distinguish only what was truly worth remembering. Anything that was not – that was not 'the work of Caesar' – in effect did not exist. It was 'out of this world', and therefore presumably not in Europe.

Aeneas had been fully schooled in humanism's new learning and it was his zeal as a born-again Roman that probably got the better of him on this occasion. But in reality his own world was much less confined than was or would be the norm among his fellow humanists, and the point of *De Europa* – as well as its sister work, *De Asia*, completed a few years later – was actually to open it up. In particular, Aeneas was keen to ensure that Germany would be brought in from the cold, and he spends a good deal of time dealing with the subject; so much so that he feels the need to apologize to his readers for testing their patience. 'By recording the cities of Saxony name by name', he explains

> we have broken the rule we had set ourselves for writing. We have done this because the ancient writers spoke sparingly of Germany, referring to German matters as though they themselves were dreaming and as though that country lay out of this world. For that reason I will be forgiven – and there are those, perhaps, who will be thankful – if, in describing the regions of Germany, so as to subject the matters to our gaze, we were somewhat long-winded, going beyond the limits of our plan.[11]

'Subjecting to our gaze' is a telling expression. It reminds us of the Countess of Pandino's 'imperious eyes' – or of 'imperial eyes' – and what is for sure is that Aeneas's aim was to 'teach his readers' eyes to see' in a way which does not differ all that much from what the *Pandineschi* sought to do.[12] In his case, however, rather than the need-not-to-know principle, it may be more correct to speak of the no-need-to-know principle. Although there were many places worth remembering, there were also many which were not and could safely remain 'out of this world'.

However it would be unfair, or incorrect, to suggest that Aeneas's ideal world was as confined as the *Pandineschi*'s. He, for one, was not afraid to venture beyond the confines of courts and estates, and he was certainly no armchair traveller. He may never have got as far as to-be-hurried-on-from Hibernia, but in the course of his short life he visited many countries which for most of his contemporaries would never be more than a name on a page, if that. He was an extraordinary man who lived through extraordinary times. His life and career coincided with the years when the Church, emerging triumphant from the so-called Babylonian Captivity (during which the popes had been long-term hostages of the French in Avignon) and the ensuing papal schism (during which there had been at first two, and then three popes all at once), and managing to defuse the threat of conciliarism (i.e. of those who fought to install a council instead of the pope as the Church's governing authority) and of Bohemia's Hussite heresy (a proto-Protestant movement), re-established itself in Rome as a universal monarchy. They were also the years when the Holy Roman empire, which, thanks to the astuteness and longevity of Frederick III (1415–52–93), was in the process of becoming the hereditary apanage of the Habsburgs, succeeded in imposing itself once again as a major player in world affairs. They were the years, as has already been mentioned, when Italy consolidated its position as the leading intellectual and artistic power of the day. But more ominously, they were the years when the Turks, seizing Constantinople and permanently establishing their capital on European soil, seemed poised to become an unstoppable and over-mighty imperializing power. In all of these events Aeneas played a leading role.

Born into an impoverished noble family from Siena, he studied law and began his working life as secretary to a cardinal, whereupon he never looked back again, turning himself from a libertine (as a youth he wrote bawdy works and led a life of dissipation) into a Church reformer and crusader (he played a leading role in the negotiations with the Hussites, and his dying act was to attempt to lead the crusade he had made it the mission of his pontificate to promote), from a conciliarist (for close on ten years he acted as secretary and minute-taker to the Council of Basel) into a papal monarchist (it was he who, in 1459, summoned the Diet of Mantua, which drove the final nail into the coffin of conciliarism), from supporting an antipope (he became secretary to Felix V, formerly Duke Amedeus VIII of Savoy, who was elected pope, in competition to Eugenius IV, by the Council of Basel in 1439) to being pope (he was Pius II from 1458 to 1464). And even as he consolidated his authority within the Church (he was appointed secretary to Pope Eugenius IV in 1446, took holy orders in 1447, became Bishop of Trieste that same year, Bishop of Siena in 1450 and a cardinal in 1456), he strove tirelessly to defend the power of the empire (from 1442 onwards he was one of the closest and most trusted advisers of Frederick III). Moreover, he was an indefatigable traveller, constantly on the move between Italy, Germany and Austria, paying visits to almost every part of Italy, sent on one occasion to Scotland, as we have seen, and on his very death-bed borne across the Appenines, riddled with gout, so as to be present in Ancona to bid farewell to the crusaders whom he had called to arms (or rather, to those few crusaders who had bothered to answer the call). He was also an amazingly prolific writer. Besides *De Europa* and *De Asia* and his autobiography, one of the first ever written, his works include an erotic short story and a bawdy comedy, poetry (for which he was crowned poet laureate by Frederick III in 1442), moral treatises (on education, on the pitfalls of courtly life), as well as histories (of the Bohemians, of Frederick III). To top it all he was a keen patron of the arts, who oversaw the transformation of his birthplace, Corsignano, into a model Renaissance town which he named Pienza, after himself, and which is one of the most delightful places to visit in the whole of Tuscany.[13]

Standing on the balcony of the palace which Aeneas built in Pienza and admiring the stunning Val d'Orcia which it overlooks (nowadays a UNESCO world heritage site), one could well understand why he felt that the face of the earth in countries untouched by the winter sun was too wild and uncivilized to be worth remembering. Other parts of the world, however, would not have been so easy to dismiss from his mind. The Turks had seen to that. In *De Europa* the 1453 fall of Constantinople is called 'that perpetual monument to the sloth and cowardice of Christians', and in *De Asia* Aeneas laments the fact that

> to such an extent has the stock of the Turks risen, that what used to be called Asia, they now call Turkey [...]. Asia believed in Christ and [...] all the filth of the old religion ceased to be [...], but with the advent of the Turks everything has changed [...]. The former corruption and shamelessness have returned, for even though the Turks hate idols and claim to believe in a single God, they nevertheless

> give credence to unworthy views on Christ and do not worship the Trinity, hardly fathoming how it is possible for God to be triune. And, slaves to pleasures and to lust, they have offended Holy Writ. Holy learning is no longer, and the study of all the noble arts is at an end.[14]

It was in pursuit of that one idea that Aeneas wrote *De Asia*, as a complement to *De Europa*: to topple the monument to Christian sloth and cowardice, and in its stead to build an Asia which would once again 'believe in Christ'.

At the same time Aeneas was just as keen that Asia, and the rest of the world, should also believe in Rome. His zeal as a born-again Christian – being a libertine turned pope, the expression is certainly apt – was matched in every way by his zeal as a born-again Roman, much of which was stoked by his familiarity with a newly rediscovered work from ancient times, the *Geography* of Strabo, which was to exert an enormous influence on the cultural and political élite of the time, and even of later centuries. Aeneas was one of the earliest readers of Strabo, and in an age when the private ownership of books could not be taken for granted, since the printing press had not yet been invented, he owned his own copy of the *Geography*, which he carefully annotated and which it is still possible to view and consult in the Vatican Library. It is only right, of course, that that is where it should be housed, for the Church and popes were amongst the keenest promoters of the study of geography. The study may not have involved works which were specifically Christian, but without such works it would not have been possible to construct a Christian world – a Holy Roman world, not to say a wholly Roman world.

* * *

Strabo (63? BC–24? AD) was a Greek-speaking Roman citizen from the Pontus region of the Black Sea, whose massive *Geography*, written in Greek, never in fact enjoyed very much popularity in his own day. It was better known in the Byzantine world (that is, the eastern, Orthodox part of the Roman empire, with its capital in Constantinople, which had continued to survive after the demise, in the fifth century AD, of the western Roman empire) and it was via Byzantium that it reached the west. Having first appeared in Italy in the 1420s, it was propelled to fame in the late 1430s, at the Council of Ferrara-Florence (a church council summoned to heal the rift between the Latin west and the Orthodox east). Soon thereafter it was translated into Latin, at the behest of Pope Nicholas V (1447–55), the founder of the Vatican Library, and within a short while it had become an indispensible handbook for the high and mighty, as Strabo had indeed hoped it would be. The nobility were to be brought up on it – one of its Latin translators (there were two) was Guarino da Verona (1374–1460), a renowned humanist educationalist, who ran a model school for the sons of the rich in the northern Italian city of Ferrara. Lords were to desire it – Lorenzo the Magnificent (1449–92), the head of the Medici household, Florence's leading family, commissioned a manuscript of the Greek version for his library, the ruler of Urbino, Federigo da Montefeltro (1422–44–82), acquired a Latin one for his own. Kings were to be gifted it – Guarino's Latin translation was offered to the deposed King of Naples, among the first to own a copy of it was the King of Hungary. And even emperors, four and a half centuries on, were to feel its fascination – Napoleon I (1769–1821) ordered it to be translated into French. Such was the demand for it in Renaissance Italy that the Latin version – in Guarino's translation for Books One to Ten and that of Gregorio Tifernas, another humanist (1414–c.1462), for Books Eleven to Seventeen – was published in Rome, in 1469, no sooner had printing been invented. Thereupon it was republished many a time, while the production of manuscript copies also continued unabated (there are at least thirty extant fifteenth-century manuscripts of it). The Greek version was published in 1516, in Venice, one of the first works ever to be printed in that language; and an Italian translation was produced and published in the 1560s, after the work had been on the market for well over a century and the world had acquired a 'face' the ancient Romans, or Aeneas for that matter, could never have imagined.[15]

Strabo's *Geography* is in long hand and provides a narrative account of the world. A synthesis of all preceding geographical knowledge, after two introductory books, which deal with theoretical questions, it leads the reader on a prolonged tour of the *oikoumene*, that is, of all the inhabited lands, beginning with the Iberian peninsula and proceeding clockwise around the Mediterranean. Its focus is on physical features (seas, rivers, mountains, etc.), but only insofar as they serve as a backdrop to the world of human beings, for what really interests Strabo is the history of human settlement, and in particular of peoples and cities organized in such a way as to allow for the full deployment of human activities and the orderly conduct of trade. Strabo's geography, in other words, is human geography, but it is also political geography, in that it is explicitly aimed at providing men involved in the affairs of state with the information they need to conduct their business, for it is not possible to rule over men, Strabo maintains, unless one knows those men and where they live. It is not just any politicians, however, Strabo had in mind, but those politicians who had given effective proof of their ability and right to rule, that is, Roman politicians. The *oikoumene*, in his view, is a world which is centred on the Mediterranean – what the Romans called *mare nostrum*, 'our sea' – and whose heart beats in Rome. His geography therefore is a geography of things that matter to people who matter, in Rome, and it is predicated upon, and in turn predicates a scale of subjective values which seek to promote and uphold, and render scientifically unchallengeable, the centrality of Rome, and by the same token, the peripherality of places and peoples which, because they are too far removed from Rome and Roman concerns, can and need have no weight, or in other words of places and peoples which are not worth remembering.

To uphold this scale of values, one of the primary tasks of the geographer is to establish boundaries: 'geography seeks to define by boundaries that section of the earth which we inhabit', writes Strabo. It is also the task of the geographer to break down the world into well-defined parts, for only what is well defined can be fully grasped, and Strabo's terms of definition in the matter were to become the standard ones for Renaissance geographers: rivers, mountains, seas, tribes, shapes and sizes. 'A country is well defined', Strabo says, 'when it is possible to define it by rivers or mountains or seas; and also by a tribe or tribes, by a size of such and such proportions, and by the shape where this is possible'. At times, though, defining is not as easy as at others. When it comes to islands, for instance, Strabo tends to attach them to the mainland to which they are closest; he can thus speak of 'the islands of Europe'. When it suits him, however, he can also suggest that islands lie in a kind of no-man's water, where they wait to be attached to whatever land has a better claim on them. That tends to be the case above all with the islands off the coast of Asia Minor, which he seems reluctant to want to cede to Asia. For example, he writes that 'in the strait between

Asia and Lesbos there are about twenty small islands', as if to say: is that Asia or is it Europe?[16]

Of one thing Strabo is sure, though, and that is that Europe is the best part of the world and therefore the one which must take precedence: 'I must begin with Europe, because it is both varied in form and admirably adapted by nature for the development of excellence in men and governments, and also because it has contributed most of its own store of good things to the other continents'. That does not mean, however, that Strabo is a zealot who will twist the facts to suit his purposes. Indeed, he is much concerned about the accuracy of his information and the plausibility of his presentation, and it is not uncommon in the *Geography* to come across passages like this: 'up to this point, then, having taken as hypothesis that the most southerly regions of India rise opposite the regions about Meroë – which many have stated and believed – I have pointed out the absurdities that result from this hypothesis'. What it does mean is that for Strabo accuracy is not enough. It is not a value in its own right, for ultimately what matters is not what the world *is* like, but what it *appears* to be like – to those who would 'subject it to their gaze', as Aeneas was to say, a gaze which, as Aeneas also demonstrated, could be fairly short-sighted. Only what was Roman did it manage to perceive; the rest was to remain 'out of this world'.[17]

By that token Ireland was bound to fare rather badly, and sure enough we find in the *Geography* not only that Ireland [*Ierne*] is left out in the cold, but that it is specifically in relation to Ireland and its vicinity that Strabo spells out his views regarding the geographer's priorities:

> the distance that should be set down for the stretch from Britain to Ierne is no longer a known quantity, nor is it known whether there are still inhabitable regions farther on, *nor need we concern ourselves about the question*, if we give heed to what has been said above. For, so far as science is concerned, it is sufficient to assume that, just as it was appropriate in the case of the southern regions to fix a limit of the habitable world by proceeding three thousand stadia south of Meroë (not indeed as though this were a very accurate limit, but as one that at least approximates accuracy), so in this case too we must reckon not more than three thousand stadia north of Britain, or only a little more, say, four thousand stadia. *And for governmental purposes there would be no advantage in knowing such countries and their inhabitants, and particularly if the people live in islands which are of such a nature that they can neither injure nor benefit us in any way because of their isolation.*[18]

Strabo's is strictly an advantage-in-knowing geography – or in other words, a worth-remembering geography – and 'people who live in islands' are more likely than others to be excluded from its ken.

Both scientifically and governmentally speaking, Ireland thus borders on the immaterial, as far as Strabo is concerned. What is more, Strabo also sees Ireland as marking the spot where civilization gives way to barbarity and humanity gives way to savagery, or rather, where civilization cannot *but* give way to barbarity and humanity cannot *but* give way to savagery. The climate will not allow it to be otherwise. Only where the 'winter sun' shines, to quote Aeneas once again, can culture thrive, and so 'Ierne [...] is the home of men who are complete savages and lead a miserable existence because of the cold'. Such indeed is their savagery that 'they are man-eaters as well as heavy eaters' and 'they count it an honourable thing, when their fathers die, to devour them and openly to have intercourse, not only with the other women, but also with their mothers and sisters'. True, to all of this Strabo hastens to add that 'I am saying this only with the understanding that I have no trustworthy witnesses for it'.[19] However, for Renaissance writers, who were in awe of him, Strabo was witness enough not to have to worry about such subtleties. What it said in the *Geography* could not but be true.

* * *

Renaissance humanists had unbounded faith in the authority of the book, especially if that book happened to be ancient. They did not regard it as a dead relic from the past; rather they considered it to be as alive and reliable as any witness from their own age. Ancient authors were teachers and friends, on whom they could always rely, and their advice was just as relevant to the present as it had been in their own day. Nobody expressed this belief more memorably than Niccolò Machiavelli (1469–1527) who, shortly after completing *The Prince*, in 1513, wrote to one of his friends:

> come evening, I return home and enter into my study; and on the doorstep I take off those everyday clothes of mine, that are full of mud and filth, and I dress up for the courts of kings; and so, decently attired, I enter into the ancient courts of ancient men where, received lovingly, I seek nourishment from that food which is mine alone and which I was born for; there I am not ashamed to speak to them and to ask them the reasons for their deeds; and they, with all their humanity, answer me; and for

> the next four hours I do not have any worries; I forget all my troubles, I do not fear poverty, death does not frighten me: all of me turns into them.[20]

The belief was shared by all humanists, even those whose main interest was geography. To find out about a foreign land, they felt, one was just as well advised to read a book as attempt to go and visit the place, or at the very least to read a book *before* going to visit the place. So – we might ask ourselves – what books were there in 'the ancient courts of ancient men', aside from Strabo's, which could speak to them of Ireland?

Above all there was another work entitled *Geography*, which had also been written by a Greek-speaking Roman citizen, and had also been forgotten during the Middle Ages, and also reached the west via Byzantium. Its author was Claudius Ptolemy (90?–168 AD), from Alexandria in Egypt, whose fame had been kept alive during the Middle Ages, thanks mainly to the Arabs of northern Africa and southern Spain, as the author of the *Almagest*, a book on astronomy. The *Geography* came to light again in the thirteenth century, and from Byzantium it reached Italy a century or so later, but it was only after Italian humanists had translated it into Latin, in the first decade of the fifteenth century, that it began to make any real impression on Europeans; and once printing had been invented and the *Geography* was able to reach an ever wider audience, its popularity and influence grew beyond measure, so much so that it has been called the 'true Bible' of the Renaissance. The first edition was published in Vicenza, in 1475, and thereafter it remained a best-seller, in Italy and the rest of Europe, until it was finally knocked off its pedestal, one hundred years later, by the maps of the German geographer Gerardus Mercator (1512–94) and the atlas of the Flemish cartographer Abraham Ortelius (1527–98).[21]

Ptolemy's *Geography*, that is, continued to be one of the standard geographical reference works throughout the period of the so-called great discoveries (the 1492 discovery of America, the 1497 circumnavigation of Africa, the 1519–21 circumnavigation of the globe, etc.), and it did so *despite* the great discoveries, for Ptolemy, of course, had not known about America, and what is more he had attached Africa to India, making of the Indian Ocean a land-locked sea. True, the contents of the work were, to an extent, updated in line with the new information which the great discoverers made available, but in any case what mattered most to Ptolemy's admirers was something else. Introducing the concept of regularly spaced parallels and regularly spaced converging meridians, with the prime meridian located in the so-called Fortunate Isles, or Islands of the Blessed (today's Canary Islands), Ptolemy had aimed to fix precisely on the globe, by latitude and longitude, the position of over 8,000 places and peoples known to him and his contemporaries. His *Geography* is thus a sort of catalogue of place-names and their coordinates.

It is preceded by a lengthy theoretical introduction, which lays down the basic principles of cosmography and chorography, and also explains the rules of map-making and how to depict the globe on a flat surface. It was soon to revolutionize the study of geography, for thanks to Ptolemy geography acquired the potential to become a scientific discipline that dealt with objectively measurable notations, to the exclusion of the symbolically charged connotations with which it had been overlaid during the Middle Ages, or indeed of the politically charged connotations with which Strabo sought to infuse it, and space could be viewed as an open and neutral continuum rather than an expanse measured in terms of divine intentions or governmental purposes. What made it possible, above all, to view space as such were the maps which had been intended as an integral part of the work. Unfortunately the ones designed by Ptolemy himself had also got lost with time, and it was to be the late thirteenth century before they were drawn, using Ptolemy's own principles, and added to the work once again. In Italy they began to circulate with the Latin translation soon after it had been carried out, but they were only to be incorporated into the second printed edition of the work, which came out in Bologna in 1477. From then on, however, every new edition of the *Geography* contained the twenty-seven so-called 'old' Ptolemaic maps (one of the world, ten of Europe, four of Africa, and twelve of Asia), to which in due course, as the world became better known and Ptolemy's limitations better appreciated, a whole series of so-called 'new' maps was added.

Like the *Geography* of Strabo, Ptolemy's work was to be of particular interest to the mighty, for it was they above all who wished to see the world 'subjected to their gaze'. Its Latin translation, for instance, which was carried out at the papal curia, was dedicated to a pope, Gregory XII

(1406–9), and after his deposition, rededicated to another, Alexander V (1409–10). In the 1460s a luxurious manuscript of it was prepared for the lord of Ferrara, Borso d'Este (1413–50–71), and at about the same time another exceptional manuscript of it was given as a prized wedding gift, by Francesco Berlinghieri, an author we shall be encountering again later on, to Lorenzo the Magnificent and the latter's young Roman bride, Clarice Orsini. And in 1478, the third printed edition of the work, which by all accounts is the most splendid of the earliest editions, was dedicated to Lorenzo's arch-rival, Pope Sixtus IV (1471–84), after whom the Sistine chapel is named. Another of its avid readers was Christopher Columbus (1451–1506) who, but for Ptolemy, might never have discovered the Indies of the west. Ptolemy, indeed, had rather underestimated the size of the globe and overestimated the extent of the landmass, making the western passage to the east, from Spain to the Indies, seem like a feasible enough proposition, and so it was that Columbus felt emboldened to set sail and chanced upon America.

Given the special appeal of geography to men of power, it is debatable whether Ptolemy did indeed free its study of symbolical and ideological conditioning. It is also debatable whether maps are neutral representations of the world. It is they after all which allowed the world to fall into the grasp of conquerors, and all maps need a point of reference, be it the Fortunate Isles (itself an interesting choice for the world's prime meridian, positioning as it does ground zero where the blessed dwell), or Greenwich (in that other dwelling place of the blessed, Albion the Brave), or Alexandria (the epicentre of the world in Ptolemy's own eye/I). Maps also need to be put in sequence, and it is certainly no coincidence that in all the editions of Ptolemy's *Geography* it is always Europe that comes first. Nevertheless, it is true to say that what the *Geography* did was to provide the tools whereby the different parts of the world could be perceived not in terms of whether or not they were worth remembering, but simply of where they lay. With its five named promontories, its six named capes, its dozen or so named towns, its fifteen named rivers and its sixteen named tribes, Ireland, or rather, as Ptolemy calls it, *Iwernia*, could now be seen to lie not 'out of this world' or beyond Rome's field of vision, but simply in the top left-hand corner of the map of the world and of the first map of Europe.

* * *

We should not forget, however, that as long as Ptolemy's *Geography* remained in vogue, so too did Strabo's. Those who liked to talk with ancient books therefore always had to contend with at least two different ways of subjecting Ireland to their gaze. In fact they had to contend with far more, for Ptolemy and Strabo were by no means the only ancient authors which they liked to keep on their bookshelves. In truth, the ancients did not have very much to say about Ireland, but given the esteem in which they were held, what they did have to say played a fundamental role in shaping, not to say confusing the vision of Ireland, and by extension of the world, which at that very time was being crafted anew, by and for Europeans. One of the most influential such works, rediscovered in the fourteenth century, was the *Natural History* by Pliny the Elder, a Roman writer of the first century AD, a 'good imperial bureaucrat', as he has been labelled, who attempted, in the *History*, to draw up an inventory, for the satisfaction of Rome's élite, of all the marvels of the world. A massive work, it caused the Renaissance to marvel too and became extremely popular. No sooner had printing been invented than it was published, in 1469, in the original, and between then and the middle of the following century no fewer than thirty-eight editions of the Latin text saw the light of day, besides the seven editions of an Italian translation carried out by Cristoforo Landino, a prominent Florentine humanist whom we shall also be meeting again later on. Regarding Ireland all Pliny had to say was that it measured 600 by 300 miles and that it was situated 'above Britain' [*super Britanniam*]; but that may well have been the cause of some bemusement for those who believed, as they did, that Ireland lay between Britain and Spain.[22]

In order to find out about Ireland's marvels, Renaissance readers would have needed to consult another work which was rediscovered in the course of the fourteenth century, the *Cosmography* or *On the State of the World* [*De situ orbis*] by Pomponius Mela (first century AD). Written in about 44 AD, the work was a cross between a school book and a handbook for the general public and its aim was to provide a systematic description of the known world, with the emphasis, once again, on marvels. The Renaissance took to these with the same enthusiasm as it did to Pliny's. The *Cosmography* was first published, in the original Latin, in 1471, another sixteen editions of it saw the light of day over the following half century, and it was translated into Italian (by one Tommaso Porcacchi, whom we shall also be meeting in due course) as late as 1557. Those who were curious therefore had plenty of chances to find out about Ireland. Not that Mela had all that much to say on the subject; but what he did, is certainly memorable. For Mela, Ireland means sweet grass, bursting cows and uncouth inhabitants:

> Above Britain – he writes – there is Ireland [*Iuuverna*], which is almost the same size, though it is oblong on account of the equal length of its coastline on either side; its climate is hostile to the ripening of crops, yet it has such an abundance of grass, which is both luxuriant and sweet, that the cattle only need a small part of the day to eat their fill and, if they are not prevented from feeding, they burst from overgrazing. Its inhabitants are uncouth and more ignorant of virtues than any other peoples (though educated up to a point), and totally devoid of piety [*pietas*].[23]

Clearly, the advantage in knowing about Mela's Ireland would have been as insignificant to the governmental purposes of Rome as would have been the advantage in knowing about Strabo's. Yet there were many who could not get enough of such marvels.

In the third century AD the Roman grammarian and compiler Solinus put together a work which came to be known variously as *Multihistory* [*Polyhistor*], or *Miscellany of Memorable Things* [*Collectanea rerum memorabilium*], or *On the State and Memorable Things of the World* [*De situ et memorabilibus orbis*]. The work is mostly a compendium, as the author himself acknowledges, of Pliny's *Natural History* and Mela's *Cosmography*, and although its pedigree was bound to make it suspect to humanist purists, because it had not been written during Rome's golden age, its popularity during the Renaissance, by when it had been in circulation almost uninterruptedly since the time it was first written, grew even greater. The first published edition, which came out in Venice in 1473, was followed by another ten Latin editions, and an Italian translation was brought out, also in Venice, in 1557. The printed editions, however, bore little resemblance to the original, especially with regard to Ireland, for such had been the fascination caused by the marvels of the *Polyhistor* that many more were added to the work over time. This is what they said about Ireland (the sentences or parts of sentences in square brackets being the later additions):

> Britain is surrounded by many known islands, of which *Hibernia* is closest to it in size. It is inhuman because of the rough manners of its inhabitants. On the other hand it is so rich in pastures, that the cattle there, unless it is intermittently kept away from its food [during the summer], is in danger of overeating. There is no snake there, only the rare bird, and the people are inhospitable and warlike. Victors, having first drunk from the blood of those they have killed, paint their faces with it. Licit and illicit is all one to them. [When a woman has given birth to a male, she places its first feed on the sword of her husband and gently introduces it into the mouth of the child on the very tip of the sword as a foretaste of things to come, and with pagan invocations she prays that he may encounter no other death save in war and amid arms. Those who care for polish decorate the hilt of their sword with the teeth of sea monsters. In that way they make them shine with the splendour of ivory, for a man's greatest glory is in his arms.] Nowhere are there any bees, and if anyone should take earth or stones from there and place them in a hive, the bees will swarm away from their combs. The sea which flows between Ireland and Britain is rough and agitated [all year round], and is not navigable except for a very few days [in the summer]. [They sail, though, in crafts made of reed, with a wrapping of cowhide. However long the crossing might take, sailors abstain from food, confident that] the full distance is no more than 120,000 strides, according to the reliable reckoning of those who have calculated it.[24]

Possibly even more popular with Renaissance readers than Solinus's *Multihistory*, even though its linguistic pedigree would have been as suspect as his, since it too had been written during the post-classical age, was *The Book of Etymologies or Origins* [*Etymologiarum sive originum libri*], otherwise known as *On the Nature of Things* [*De rerum natura*], by the Spanish Church Father Isidore of Seville (570?–636).

A massive encyclopaedia, which owed a lot to Solinus and had remained a standard reference work throughout the Middle Ages, *The Book of Etymologies* is organized according to topics, and one of its sections deals specifically with islands, of which, Isidore explains, only the most famous and the biggest are recorded. Ireland is one of the very first islands to get a mention, and though the entry is short, and possibly confusing, it came to acquire almost proverbial status. 'Scotland', says Isidore,

> like Ireland [*Hibernia*] is an island close to Britain which is smaller in extent but more fertile on account of its location. The latter extends from south to north, and its nearest parts look towards Iberia and the Cantabrian Sea, wherefore it is called Hibernia. But it is also named Scotland, because it is inhabited by Scottish peoples. There there is no snake,

> only the rare bird, and no bee, so that, if anyone should take earth or stones from there and place them in a hive somewhere else, the bees will swarm away from their combs.[25]

Though Ireland, on the reckoning of Isidore, did deserve a place among the most famous of islands, what made it famous, once again, was essentially its marvels.

All told, what the talking books from the 'ancient courts of ancient men' marvelled at, regarding Ireland, was what it had (the cold that turns its inhabitants into complete savages, the sweet grass that causes cows to burst, the earth and stones that frighten bees) just as much as what it did not have (crops, snakes, bees, birds, piety and virtue). What confused them was its size (same size as Britain? smaller than Britain?), its location (above Britain? between Britain and Spain?) and its name (*Ierne*? *Iwernia*? *Hibernia*? *Scotia*?), not to mention the identity of its inhabitants (Scottish? with a touch of Spanish?). And what caused them to look askance at it was those very inhabitants: inhuman, inhospitable and incestuous cannibals, who could not tell the difference between licit and illicit. However, they did recognize that the Irish were warlike and that some of them at least had it in them to be victors; and that was the surest sign that a people may not be destined forever to be left 'out of this world'. Human history, according to the ancients, was essentially the history of wars, and peoples mattered more or mattered less depending on whether they could win or not. Ultimately, of course, victory would fall to only one of them – the Romans – but what was of concern for governmental purposes was nonetheless the degree to which others – non-Romans, barbarians – could 'benefit' or 'injure'.

* * *

In truth hardly any of the ancient authors considered Ireland from that point of view. The only one who did was the historian Cornelius Tacitus (56?–118 AD), who, in his biography of his father-in-law, the Roman general Agricola, penned what has been called, somewhat exaggeratedly perhaps, 'one of the most important and informative passages on Ireland in classical literature'.[26] The passage is actually quite short, but there is no doubt that what it lacks in substance it carries in weight, for Tacitus was one of the ancient writers whom Renaissance humanists, not least Aeneas Silvius

Piccolomini, admired most. Agricola, says Tacitus, having invaded Britain and led his troops as far as the island of *Mona* [Anglesea], off the north-western tip of Wales, then planned to cross over to Ireland, reckoning that it might be of benefit to 'the arms of Rome' if he did so. He thus

> garrisoned the coast of Britain facing Ireland [*Hibernia*], motivated more by hope than by fear: for, since Ireland lay half-way between Britain and Spain, and was within easy reach of the Gallic sea, he felt that it might be used to unite the best recruiting grounds of the empire with great mutual advantage. Compared with Britain, Ireland is small in size, but larger than the islands of our seas. In soil and climate, and in the temperament and manners of its people, it differs little from Britain; its coasts and harbours are better known, thanks to trade and to merchants. Finding that one of the native kinglets [*reguli*] had been driven from his throne by family dissension, Agricola had welcomed him under a show of friendship, and now preserved him for future use. I have often heard him say that a single legion with a moderate force of irregulars could conquer and hold Ireland. Such a conquest would, he considered, be of great service against Britain. The arms of Rome would stretch to the horizon and liberty would be swept out of sight.[27]

In the end, Agricola did not carry out his plan and therefore Ireland was to have no part to play in the forward march of Rome's *imperium*, but what was not to be lost on the would-be empire-builders of the Renaissance and their humanist apologists was the spirit of Agricola's intentions, which gave legitimacy to their own governmental purposes. They too (or some of them, at least), as they themselves sought to 'stretch to the horizon' the arms of Rome born-again and so to 'sweep liberty out of sight', would wonder whether or not Ireland could be a 'recruiting ground of empire', and whether or not there was any 'use' in Irish 'friendship'. Above all, they would take satisfaction in the knowledge that Irish rulers were no more than 'kinglets'.

They would also take comfort in the view of history which *Agricola* propounds. There can only be history, the work makes clear, where there are conquerors. 'Many writers', Tacitus says,

> have written of the site and peoples of Britain. I shall deal with the subject, not to challenge comparison with their accuracy or insight, but simply because the island was now for the first time *completely conquered*. Where others have embellished their ignorance with rainbow rhetoric, my story will rely *simply upon facts*.

Only when it is imperial, that is, can history be said to be factual and true. The enemies of empire, for their part, cannot have a history, for they are ignorant barbarians, who express themselves in 'discordant cries' and do not know what they talk about: 'whether the original inhabitants of Britain were indigenous or invaders has never been determined, as is not unusual among barbarians'.[28] In other words, to be worth remembering, barbarians must inhabit a world which not only has one, recognizable face – a face that is recognizably Roman – but a world which speaks with one voice – the right voice. Otherwise they are doomed to remain out of 'this' world: outside the written world, beyond the written word.

That beyond, however, could not be allowed to be written off altogether. On the one hand it had to stand as a reminder of how much greater 'this' world could still become. The ancients called it 'furthest Thule' [*ultima Thule*] – a mysteriously distant island, somewhere to the north west, at the very edge of the world – and together with the poet Virgil (70–19 BC) they dreamt of it as the place where their imperial destiny would become manifest in the fullness of its glory: 'Oh would but furthest Thule be thy servant', Virgil sings to Emperor Augustus.[29] On the other hand the beyond had to serve as a warning that encountering that destiny should not be taken for granted; and that is how it is portrayed in another work by Tacitus to have enjoyed great popularity among Renaissance humanists, his *Germania*. An ethnographic survey which describes, first collectively then singly, the various German tribes, overall *Germania* paints a sympathetic portrait of the Germans. Tacitus approves of their simplicity, their resistance to the corrupting effect of civilization. 'In Germany', he claims, 'no one laughs at vice, nor calls mutual corruption "the spirit of the age"'.[30] More importantly, he marvels at the invincibility that simplicity procures. The Romans, he notes, had been fighting the Germans for over two hundred years, but still the Germans kept their independence.

It is more than likely that Tacitus's simple German was conceived as a wake-up call for the Romans, and in particular for the Roman faction which opposed the interests of his own family. Beware – he wished to warn them – there are now greater virtues in savages than in you, and soon the arms of Rome will no longer be able to stretch to the horizon. At the same time, however, the work is definitely subversive, and it has rightly been claimed that the German of *Germania* is the great-grandfather of the noble savage, that figment of a western imagination tired of its own excesses and which fancied it could perceive in the savages it had once derided the nobility of human sentiments untainted by the vices of what it had believed was civilization. *Germania*, that is, gives a voice, albeit in falsetto, to the once barbarian Other, and it demonstrates that designing foreign lands need not necessarily mean having designs upon them. Otherness could be appraised without being despised.[31] The lesson would not be lost on Renaissance humanists, even though they might have felt, like Aeneas, that it was preferable, all told, to assimilate the Germans rather than admire them for their noble otherness. Or better still, that it was preferable to eradicate them – for was that not what they were taught to do by the 'work of Caesar'?

In *The Gallic War* [*De bello gallico*] Julius Caesar (102?–44 BC), with no little amount of self-satisfaction and wounded pride, for he is very deft at giving the impression that it is the conquerors who are the wronged party, tells the story of his victory over the Gauls and the Germans. The work consists for the most part of war plans, battle scenes and harangues, which humanist historians then came to regard as the canonical format in which to couch a history; but it also contains a number of geographical and ethnographical descriptions, and these too were to serve as a model for humanist historiographers, all the more so as Cicero (106–43 BC), one of the humanists' other great masters, had pronounced in his *Orator* that 'the nature [of history] requires chronological arrangement and geographical representation'.[32] On the one hand these geographical representations, as indeed those of other ancient writers, acted as a stylistic model. They were presented and seen as digressions from the main topic of the work, and digressions, Caesar's imitators realized, were a necessary feature of a well-written history, that added to the readers' enjoyment and helped to keep their curiosity aroused. On the other hand the descriptions helped to strengthen the ideological presumptions on which the work was based. In every case they are descriptions of peoples Caesar will or would like to defeat, and while conveying the message that it is necessary to know one's enemies, they also stand as markers of Otherness ('these people are *different* from us') and by implication of the inevitability of the eradication of Otherness ('it is *because* they are different that they shall be defeated and subjected to us').

However, like Strabo, Caesar also worked according to the principle that some barbarians are more unequal than others, and consequently the eradication of Irish Otherness had no real part to play in his designs. All he says about Ireland – which he is the first to call *Hibernia*, thus bequeathing the name to posterity – is that it is 'smaller by half, as it is thought, than Britain', that it lies between Britain and Spain, and that it is as far away from Britain as Britain is from Gaul.[33] Caesar, of course, never got as far as Agricola would do, so it makes every sense that he should not have written more about Ireland. Nevertheless, one should not fail to note that for Caesar too, as for Strabo, objective distances told no more than half the story. In the Renaissance, as we now know, that half was told by Ptolemy and those who, like him, aspired to make the 'writing' [*graphy*] of the 'world' [*ge*] mimetic, whose aim, that is, was to produce maps. But there also was the other half, told by Strabo and those who, like him, aspired to make geography discursive and rhetorical. Their aim was to describe the world. Yet the aim, at least initially, was never to describe it all. For those geographers distances were subjective. They were assessed in terms of the matters-more-or-matters-less logic (the need-to-know/no-need-to-know logic) which made it necessary to ask: what peoples and lands – what islands – are out of 'this' world? Are peoples and lands, and in particular islands, Christian and Roman enough to have a history? Is Ireland worth remembering? And if it is, *which* Ireland is it that is to be remembered? The Ireland of marvels or the Ireland of governmental purposes?

Notes

1. In one way or another (i.e. partially or wholly), all the ancient works listed here were rediscovered in the fourteenth or fifteenth century.
2. 'QVE SVB FRIDERICO, tertio eius nominis imperatore, apud Europeos et, qui nomine christiano censentur, insulares homines gesta feruntur memoratu digna mihique cognita, tradere posteris quam breuissime libet' (PIUS II (2001), p. 27: emphasis added). On the emergence of the idea of Europe see Chabod

(1971), Hay (1957), Pagden (2002). For a more detailed analysis of Pius's view of Europe see Haywood (2007).

3. On Ireland's saints and scholars see Berardis (1950), Bieler (1963), Löwe (1982), Scherman (1981), Tommasini (1937).
4. On European pilgrims to St Patrick's Purgatory see Haren and Pontfarcy (1988) and below Chapter Four.
5. 'Erano in quello tempo pochi cristiani per lo mondo, almeno in Europia, perché di nuovo s'era battezzato Gostantino e la città di Roma. Ed erano poco tempo inanzi stati in Brettagna e in Irlanda baroni e prencipi de' cristiani, perché v'era stato lo re Uterpandragone e 'l re Artù con molta bella compagnia; [...] e quegli d'Inghilterra si convertirono eglino a fede cristiana, sì che in tutte le parti di ponente era Inghilterra e Irlanda fatti cristiani. [...] Tutte l'altre province erano saraini e pagani: Spagna, Francia, Borgogna, Gesina. In Asia erano cominciati verso l'India e verso Ermenia molti cristiani, e già in oriente cominciavano. E perché Gostantino si era battezzato, aveva molti nimici; e tutte l'altre fedi lo cominciarono a odiare' (BARBERINO (1969), p. 16).
6. GERALD OF WALES (1982), pp. 102, 106. 'Gens igitur haec gens barbara, et vere barbara' – 'gens omnium gentium in fidei rudimentis incultissima. Nondum enim decimas vel primitias solvunt; nondum matrimonia contrahunt; non incestus vitant; non ecclesiam Dei cum debita reverentia frequentant. Quinimmo, quod detestabile valde est, et non tantum fidei sed et cuilibet honestati valde contrarium, fratres, pluribus per Hiberniam locis, fratrum defunctorum uxores non dico ducunt, sed traducunt' (GERALD OF WALES (1964), pp. 152, 164). For more on Gerald of Wales see next chapter.
7. 'intellexit [...] non ad homines se, sed ad bestias destinatum [...] christiani nomine, re pagani. Non decimas, non primitias dare, non legitima inire coniugia, non facere confessiones; paenitentias nec qui peteret, nec qui daret, penitus inveniri. Ministri altaris pauci admodum erant' (BERNARD (1963), p. 325).
8. 'Hibernia nunc nobis absoluenda esset, que paruo a Britania disiuncta freto, partim libera Scotorum amicitia societateque gaudet, partim anglicano paret imperio. sed quoniam nihil dignum memoria per hoc tempus, de quo scriptio est, gestum accepimus, ad res hispanicas festinamus' (PIUS II (2001), pp. 186–7).
9. 'fui septem mensibus extra terrarum orbem in regno Scotie' (quoted in PIUS II (2001), p. 185); 'terra frigida' – 'Vulgus, pauper et incultum' – 'dicere [...] solitus erat Aeneas' – 'At cum eo venisset miraculumque cupidus investigaret, comperisse id mendacii; seu verum est ultra in Orchades insulas relegatum' – 'quod Caesaris opus dicunt. Ibi primum figuram orbis et habitabilem terrae faciem visus est revisere. Nam terra Scotia, et Angliae pars vicina Scotis, nihil simile nostrae habitationis habet: horrida, inculta atque hyemali sole inaccessa' (PIUS II (1984), pp. 22–8).

10. Rubiés (2007): 'Travel Writing and Ethnography', p. 14.
11. 'Excessimus scribendi modum Saxonie ciuitates nominatim commemorantes. id fecimus, quia ueteres scriptores parcissime de Germania locuti sunt et, tamquam extra orbem ea natio iaceret, somniantes quodammodo res germanicas attingunt. ob eam rem dabitur mihi uenia (fortasse et aliquis gratiam habebit), si germanicas describentes prouincias, ut res oculis subiciamus, paulo prolixiores fuerimus, propositi nostri metas egressi' (Pius II (2001), p. 133).
12. Rubiés (2007): 'Instructions for Travellers: Teaching the Eye to See'.
13. For more on Pius see Baldi (2006), Mitchell (1962), O'Brien (2009), Pellegrini (2000), Secchi Tarugi (2007).
14. 'Christianorum desidie atque ingnauie memoriale perpetuum' (Pius II (2001), pp. 246–7). 'Adeo Turcarum nomen auctum est, ut quae olim Asia uocabatur, nunc Turciam uocitent. à Turcis enim uniuersa possidetur [...] credidit Asia Christo, & [...] cessauit omnis spurcitia ueteris cultus [...] uerum Turcarum aduentu cuncta immutata sunt. [...] Redijt foeditas & pristina turpitudo. nam quamuis simulacra horrent Turcae, & unum se colere Deum dicunt, indigna tamen de Christo sentiunt, nec trinitatem adorant, haud intelligentes quomodo possit Deus unus & trinus esse, & scripturam sacram deprauarunt, uoluptatis ac libidinis serui. Perierunt sacrae literae, & omnia bonarum artium studia defecerunt' (Pius II (1571?), p. 385).
15. On Strabo and his popularity in the Renaissance see Aujac (1993), Broc (1996), Diller (1975), Freeman (2001), Milanesi (1992), Pastore Stocchi (1992), Prontera (1992), Salmeri (1988).
16. Strabo (1988–97), vol.1, pp. 371, 315, 492–3; vol. 6, p. 147.
17. Strabo (1988–97), vol. 1, pp. 485–7, 289, 503–5.
18. Strabo (1988–97), vol. 1, pp. 444–5 (emphasis added).
19. Strabo (1988–97), vol. 1, p. 443; vol. 2, pp. 259–61.
20. 'Venuta la sera, mi ritorno in casa ed entro nel mio scrittoio; e in su l'uscio mi spoglio quella veste cotidiana, piena di fango e di loto, e mi metto panni reali e curiali; e rivestito condecentemente entro nelle antique corti delli antiqui uomini, dove, da loro ricevuto amorevolmente, mi pasco di quel cibo, che *solum* è mio, e che io nacqui per lui; dove io non mi vergogno parlare con loro e domandarli della ragione delle loro azioni: e quelli per loro umanità mi rispondono; e non sento per quattro ore di tempo alcuna noia, sdimentico ogni affanno, non temo la povertà, non mi sbigottisce la morte: tutto mi transferisco in loro' (letter to Francesco Vettori, 10 Dec. 1513: Machiavelli (1976), p. 23).
21. For a sample of Ptolemy see Ptolemy (2000), including bibliography. On Ptolemy in the Renaissance see Aujac (1993), Broc (1996: 'true Bible', p. 5), Brotton (1997), Campbell (1987), Freeman (2001), Gautier Dalché (2009), Gentile

(2003), Grafton Shelford and Siraisi (1992), Milanesi (1992), Pastore Stocchi (1992), Prontera (1992), Quondam (1985), Randles (1994). On Mercator see Crane (2003). On Ortelius see Brotton (1997), Karrow (1998), Quondam (1985); also, with specific reference to Ireland, Morgan (1999), p. 29.

22. PLINY (1942), p. 199. On Pliny in the Renaissance see Broc (1996), Freeman (2001), Pagden (1993: p. 59), Prontera (1992: 'buon funzionario dell'impero', p. 285).

23. 'Super Britanniam Iuuerna est paene par spatio, sed utrimque aequali tractu litorum oblonga, caeli ad maturanda semina iniqui, uerum adeo luxuriosa herbis non laetis modo sed etiam dulcibus, ut se exigua parte diei pecora impleant et, nisi pabulo prohibeantur, diutius pasta dissilant. Cultores eius inconditi sunt et omnium virtutium ignari <magis> quam aliae gentes [aliquatenus tamen gnari], pietatis admodum expertes' (MELA (1988), p. 82). <...> = editor's interpolation; [...] = later, post-classical addition to the text. On Mela see Aujac (1975), Aujac (1993), Freeman (2001), Prontera (1992).

24. '(Brittania insula) multis insulis nec ignobilibus circumdatur. quarum Hibernia ei proximat magnitudine, inhumana [*est*] incolarum ritu aspero, alias ita pabulosa, ut pecua[*ria*] [*ibi*], nisi interdum [*estate*] a pastibus arceantur, ad [~~*ad*~~; *in*] periculum agat satias. illic nullus anguis, avis rara, gens inhospita et bellicosa. sanguine interemptorum hausto prius victores vultus suos oblinunt. fas ac nefas eodem loco [~~*loco*~~; *animo*] ducunt. [*Puerpera si quando marem edidit, primos cibos gladio imponit mariti inque os parvuli summo mucrone auspicium alimentorum leviter infert et gentilibus votis optat, non aliter quam in bello et inter arma mortem oppetat. Qui student cultu dentibus marinantium belluarum insigniunt ensium capulos. Caudicant enim ad eburneam claritatem, nam precipua viris gloria est in tela.*] apis nusquam: advectum inde pulverem seu lapillos si quis sparserit inter alvearia, examina favos deserent. sed [~~*sed*~~] mare quod inter hanc [~~*inter hanc*~~; *Hyberniam*] et Brittanniam interluit undosum inquietumque [~~*inquietumque*~~; *in quemcumque*] toto in anno nonnisi [*aestivis*] pauculis diebus est navigabile idque [~~*idque*~~; *Navigant autem vimineis alveis, quos circumdant ambitione tergorum bubalorum. Quantocumque tempore cursus tenebit navigantes escis abstinent, freti latitudinem*] in centum viginti milia passuum latitudinis [~~*latitudinis*~~] diffundi qui [~~*qui*~~; *quod*] fidem ad verum ratiocinati [~~*ratiocinati*~~; *roncinanti*] sunt aestimarunt' (SOLINUS (1958), pp. 100–1, [SOLINUS (1512), p. XVIIr]). On Solinus see Freeman (2001), Pagden (1993).

25. 'Scotia idem et Hibernia proxima Brittaniae insula, spatio terrarum angustior, sed situ fecundior. Haec ab Africo in Boream porrigitur. Cuius partes priores Hiberiam et Cantabricum Oceanum intendunt, unde et Hibernia dicta: Scotia autem, quod ab Scotorum gentibus colitur, appellata. Illic nulla anguis, avis rara,

apis nulla, adeo ut advectos inde pulveres seu lapillos si quis alibi sparserit inter alvaria, examina favos deserant' (ISIDORE (1966), XIV, vi, 6). On Isidore see Baldacci (1992).

26. Freeman (2001), p. 56.
27. 'Eamque partem Britanniae quae Hiberniam aspicit copiis instruxit, in spem magis quam ob formidinem, si quidem Hibernia medio inter Britanniam atque Hispaniam sita et Gallico quoque mari opportuna valentissimam imperii partem magnis in vicem usibus miscuerit. spatium eius, si Britanniae comparetur, angustius nostri maris insulas superat. solum caelumque et ingenia cultusque hominum haud multum a Britannia differunt; [in melius] aditus portusque per commercia et negotiatores cogniti. Agricola expulsum seditione domestica unum ex regulis gentis exceperat ac specie amicitiae in occasionem retinebat. saepe ex eo audivi legione una et modicis auxiliis debellari obtinerique Hiberniam posse; idque etiam adversus Britanniam profuturum, si Romana ubique arma et velut e conspectu libertas tolleretur' (TACITUS (1967), p. 108). On the Renaissance rediscovery of Tacitus see Grafton Shelford and Siraisi (1992: p. 43), Kraye (1996: *passim*).
28. 'Britanniae situm populosque multis scriptoribus memoratos non in comparationem curae ingeniive referam, sed quia tum primum perdomita est. ita quae priores nondum comperta eloquentia percoluere, rerum fide tradentur' – 'Ceterum Britanniam qui mortales initio coluerint, indigenae an advecti, ut inter barbaros, parum compertum' (TACITUS (1967), pp. 98–9: emphasis added).
29. 'Tibi serviat ultima Thule' (*Georgics*, I, 30: VIRGIL (1998), p. 82). On Thule see Mund Dopchie (2009).
30. 'Nemo enim illic vitia ridet, nec corrumpere et corrumpi saeculum vocatur' – 'tam diu Germania vincitur' (TACITUS (1938), 19.3, 37.2).
31. See Grafton Shelford and Siraisi (1992: 'great-grandfather', p. 43), Jones (1971).
32. 'Rerum ratio ordinem temporum desiderat, regionum descriptionem' (CICERO (1967), p. 244).
33. 'Hibernia, dimidio minor, ut aestimatur, quam Britannia' (CAESAR (1970), p. 250).

CHAPTER TWO

Amazing World? Imagining Ireland with (and without) Gerald of Wales

Authors and works discussed in this chapter (in chronological order):

- Horace (65–8 BC), *The Art of Poetry* [*Ars poetica*]
- *Voyage of St Brendan* [*Navigatio Sancti Brendani*] (tenth century)
- Gerald of Wales, aka Giraldus Cambrensis, aka Gerald de Barri (1146?–1223), *The Topography of Ireland* [*Topographia hibernica*], aka *On the Marvels of Ireland* [*De mirabilibus Hybernie*] – also see Chapter One
- Francesco Petrarca, aka Francis Petrarch (1304–74), manuscript copy of Gerald of Wales, *On the Marvels of Ireland/Topography of Ireland*; *Letters between Friends* [*Epistolae familiares*, aka *Rerum familiarium libri*]; *On the Life of Solitude* [*De vita solitaria*]
- Giovanni Boccaccio (1313–75), *Decameron*; *Of Mountains, Woods, Springs, Lakes, Rivers, Pools or Ponds, and of the Various Names of the Sea* [*De montibus, silvis, fontibus, lacubus, fluminibus, stagnis seu paludibus et de diversis nominibus maris*]
- Domenico Silvestri (1335?–1411?), *On Islands* [*De insulis*]
- Domenico Bandini, aka Domenico di Bandino (d. 1418), *The Source of Memorable Things in the Universe* [*Fons memorabilium universi*]
- Fra Mauro (d. 1459), Borgia world map; Correr world map

For some there was no doubt about the matter. Ireland *was* worth remembering, and it was its marvels that made it so. The matter, however, was not that simple. Even though the ancients stood as guarantors for them, it was nevertheless imperative to establish whether those marvels were believable. And if they were, whether an author who aspired to be taken seriously could afford to write about them without losing face, for as Aeneas Silvius Piccolomini was to state categorically, wonders were fables and fables were trifles. Truth and seriousness were something else altogether. So in order to establish their *bona fides*, writers who believed in the fables of Ireland – or better perhaps, who wanted or needed to believe in them – had to engage in complex language-games, in which, as Petrarch memorably stated, the foot competed with the mind. The foot would take one to the place – though in reality no one's feet actually did so: the armchair was generally as far as they got – and by weighing up the evidence, the mind would forestall it, for you can only travel to places you already know about or have heard about.

Weighing up the evidence was itself a complex matter. It involved not only questioning eye-witnesses, if any could be found or conjured up, but above all 'talking with books', 'entering into the ancient courts of ancient men', and even, when the 'amazing to say' threatened to defy logic, cross-examining God, or his agents. In addition there was the problem of time. Time tampered with books and with the language they were written in, so how could one be sure that what books said was actually what they meant to say? The problem is one that faces us too of course. As we will see, our own conversations with (Renaissance) books can at times be rendered halting by the state of preservation of those books. At least humanists had the advantage that more often than not the truth they sought to establish was a truth *for*, and since they mostly wrote to convert and to educate, what their instincts told them was less to find out what Ireland was like, than what was in it for them and for their readers. Caring as they did for the Self, what they needed above all was an Ireland that was good-for (themselves). Nevertheless, a more objective kind of truth – a need-to-know-everything kind of truth, a truth *about* – was also in the making at the time, for it is arguably the case, paradoxical though it may seem, that humanists were the forerunners of modern science, who, even as they cared for the Self, attempted to establish 'the actual truth of things' on the basis of observed facts and clear evidence. That paradox, and that conflict of truths, are evident, up to a point, in all the works discussed in this book, but nowhere more so, possibly, than in the works that are the subject of this chapter.

At precisely the time Aeneas was remodelling the face of Europe, by casting 'out of this world' what was not 'worth remembering' and 'subjecting to the gaze' of his readers lands which the 'dreamers' of the past had themselves not considered to be worthy of inclusion, in Venice Fra Mauro (d. 1459), a monk from the monastery of San Michele on the island of Murano, was designing an exceptional map of the world, where the only limits were those imposed by lack of information, and where Ireland was afforded an exceptional status. Beautifully drawn and painted in brilliant colours, Fra Mauro's is a circular map, about two metres in diameter, which presents the world upside down, so to speak, with the north to the bottom and the east to the left, as had been the custom for many centuries, and which shows the three known continents of Europe, Asia and Africa, and the seas that lie between them, surrounded by a narrow strip of the ocean which was believed to cover the rest of the globe. In addition, in each of the corners, between the map itself and the frame in which it is set, there are fair-sized illustrations, together with explanatory texts, which depict, in the top left-hand corner, the heavens and the stars, in the top right-hand corner tides and continental landmasses, in the bottom right-hand corner the elements and the habitability of the southern regions, and in the bottom left-hand corner the earthly paradise. Begun in the 1440s, but still being touched up some months after Fra Mauro's death, the map originally stood in the church of the monastery. In due course it was relegated to a side room, where it remained until 1811, when Napoleon I ordered the suppression of Italian monasteries, whereupon it was moved to the Marciana Library, beside St Mark's Square, and subsequently to the Correr Museum next door, where it is once again on public view.

Its maker was a monk with a mind more practical, it would seem, than contemplative, and interests more worldly than spiritual. He acted as consultant engineer to the Venetian government, on issues relating to the conservation of the lagoon in which Venice was built and the deviation of the river Brenta, in Venice's hinterland, where many of the city's merchant princes were turning the profits of trade into landed estates and sumptuous villas; but above all he was an expert cartographer, whose renown extended far and wide, with earth-shattering repercussions, quite literally. A copy of his world map, which had been intended primarily for the government and merchants of Venice, was sent, upon request, to the king of Portugal and came to play a significant part – or so it is said – in the decision of the Portuguese to attempt the circumnavigation of Africa. Notwithstanding Ptolemy's teaching on the matter and long before the old Ptolemaic maps were superseded by the new ones, Fra Mauro's map, indeed, was the first modern map to show that Africa and India were not attached to one another and that the Indian Ocean met up with the Atlantic.[1]

What makes the map even more remarkable is that it is a talking map, so to speak, which contains close on 3,000 captions, by means of which Fra Mauro engages with his viewers, challenging them, as though in defiance of Aeneas, not to look at the face of the world through myopic lenses. Since his target viewers are practical men, used to ploughing far-off seas and travelling through distant lands, when he can, Fra Mauro aims to provide them with practical information, which will assist them on their journeys. For instance, concerning Yemen he explains: 'Here one pays a toll'. In the Indian ocean he warns: 'This whirlpool below here is a great danger for sailors'. With regard to Iceland he alerts: 'In this place live bad men and they are not Christians'.[2] But it is not just the dangers of travelling which he brings to their attention. Often he casts a more covetous or enchanted look at the world, informing them, for example, that Taprobane (usually the name given to modern-day Sri Lanka, but in Fra Mauro's case Sumatra) is

> a most noble island which, it is said, is more than 4,000 miles in its round and is divided into four kingdoms, in which one finds much gold, pepper, cloves, aloe wood, which is called *galambach* and has the finest scent in the world and is sold for the weight of gold. And here is born on trees a fruit called *durian*, which is the size of an average watermelon and has a green and knobbly skin like a pine-cone and inside has five fruits, each one the size of an average pine-cone and each one of these five fruits has its own sweet taste, and inside they are purplish in colour and they are very warm. Here one finds many elephants, and the men of this island are of better condition than those of the other neighbouring islands and they are handsome men, strong and fine and good astrologers but idolatrous. They are also taller than those who are born in India and their elephants bigger than the Indian ones, and the Indian ones bigger than those that are born in the Mauritanias, and this is because of its excellent location and the goodness of its air.[3]

At times the enchantment becomes amazement, as Fra Mauro mentions the world's many 'marvels' [*meraveie*] or 'novelties' [*novità*], and pauses to highlight the bizarre or the untoward, as he does, for instance, on the 'island of Sochotra, placed between the Sea of Persia and Aden but more towards the Red Sea and placed opposite the mouth of the said sea', whose inhabitants, we are told, 'used to be Christians and are necromancers and with that art they take away and sell favourable winds to sailors'.[4]

At other times, however, the enchantment is stretched to incredulity, and Fra Mauro seems uncertain whether or not to report what he does, knowing full well that his viewers too will find it hard to believe what they see and hear. But he tries as best as he can to reassure them:

> I think that many will be amazed because I put the source of the Nile here, but for sure if they proceed with reason and choose to investigate what I did – and with what amount of diligence I cannot say! – they will see that I have proceeded to demonstrate this on the basis of very clear evidence which I got.[5]

Like Strabo in other words (whom he may have known about, although we cannot be entirely sure) Fra Mauro suggests constructing our image of the world on the basis of reasonable conjecture, comparing what has been 'left in writings' with what people 'worthy of trust' have 'seen with their own eyes'. It is thus that he comes to the conclusion that Africa is circumnavigable:

> Many opinions and readings can be found according to which water does not surround this our habitable and temperate zone, but there are many witnesses to the opposite effect and especially those whom the King of Portugal's Majesty sent with their caravels to look for and see with their own eyes, and who say that they followed the coasts south west for 2,000 miles beyond the Strait of Gibraltar [...]. And they have made new maps of that navigation, giving new names to rivers, gulfs, capes, ports, and I have had a copy of them. Wherefore, if one insists on contradicting these people, who have seen things with their own eyes, all the more so must one not agree with nor believe those who left in their writings what they had never seen with their own eyes, but only reckoned to be so. Moreover I have spoken with a person worthy of trust, who insisted that he had been carried by a raging storm for forty days in a ship from India that crossed towards the west southwest, beyond the Cape of Soffala and the Green Islands and, according to the astrologers who are their guides, that covered almost 2,000 miles [...]. What is more Pomponius Mela says, in the third book of his *Cosmography*, that someone called Eudoxus, fleeing from Lathimin, king of Alexandria, went out of the Arabian Gulf and sailed through those southern regions and came all the way to Gades, which is the Strait of Gibraltar. Therefore, without any doubt, one can affirm that this southern and south western part is navigable, and

> that the Indian Sea is an ocean and not a stagnant sea, and that is what is affirmed by all those who sail that sea and live in those islands.[6]

Not always, however, is it easy to cast doubt aside and take evidence at face value.

What puzzles Fra Mauro most is what names to give to things and places, because

> the variety of interpreters causes many errors in the exposition of names. That is why I am not fully persuaded that one can demonstrate the truth regarding names, for it is not possible to bring about agreement with such a variety of languages, which change and confuse names according to their manner of speech, so that names are bound to be modified by languages and by the length of time and by the errors of librarians.[7]

For his good luck, however, where 'librarians' let him down, men of religion may come to his rescue. Coptic bishops had taken part in the Council of Ferrara-Florence, and there they had been questioned on the darkest mysteries of Africa. Fra Mauro, in some way, had been party to the questioning, but where others had not wanted to believe them, because they disagreed with Ptolemy, he had seen no reason not to, and on his map he thus rebuffed the doubters:

> Because to some it will seem a strange thing that I speak of this southern part, which is almost unknown to the ancients, I will nevertheless answer that this whole design, from Sayto upwards, I got from those who themselves were born there, and who are religious, and who with their own hand drew for me all these regions and cities and rivers and mountains, with their names, which I have not all been able to put in proper order for lack of space.[8]

Even men of religion, though, have their limitations, and at times Fra Mauro has no option but to appeal to a higher authority, to whom the world presents no mystery at all: 'Therefore to God Everlasting I leave the measure of his work, which he alone understands from point to point'. But God too may sometimes let him down, and when all else fails, the best, he therefore suggests, is simply to shut up: 'in these [islands] there are many novelties which we will pass under silence', 'if what I heard is true [...] I do not dare affirm'. The pen of the cosmographer should not be let to run amok. 'This noble and magnificent city – we are told, in China – if one were to narrate its refinements and its fine customs it would seem incredible and therefore it will remain inside the pen'.[9]

Nevertheless that pen can make mistakes, and so it must be afforded a second chance. Fra Mauro's map as it stands today in the Correr Museum is in fact a second draft. The first draft is now in the Vatican Library. It is known as the Borgia Map and has been called the 'dress rehearsal' for the Venice Correr Map and the map commissioned by the king of Portugal. It contains far fewer captions than the later version, and some of those are quite different from the final ones. Ireland, for instance, was to be given quite a make-over in time for the *première*. At the dress rehearsal stage, this is what Fra Mauro had said about it:

> Note that in ancient times on this island of Hibernia [*Ibernia*], which is fertile beyond measure, there was said to be a water in which whoever puts wood in it, that part which is in the ground with time becomes iron, and that part which is surrounded by water becomes stone, and that part which is above the water remains wood, and this is absolutely certain for having been seen. What is more, in one of these small islands roundabout here, all those who are born in the said island are born with a tail like that of a goat and whenever their women want to give birth, they go outside the said island and there they give birth according to their ancient and traditional customs.[10]

For the *première*, on the other hand, the carnivalesque creatures have been replaced by a cast of much more worthy characters:

> On this island of Hibernia, which is fertile beyond measure, there is said to be a water in which whoever puts wood in it, that part which is in the ground with time becomes iron, and the one which is surrounded by water becomes stone, and the one which is above the water remains wood, and if this is believed, one can also believe in the lake of Andaman; and those who wish to have full knowledge of these wondrous and other monstrous things should read Julius Solinus Polyhistor, Pomponius Mela, St Augustine, Albertus Magnus, St Thomas Aquinas in the book against the curious, as well as the *Meteorology* of Aristotle, and Pliny on the marvels of the world, and they will see that of a thousand things, I have not said one.[11]

However, it is not just the change of cast – Solinus, Pomponius Mela, St Augustine etc. instead of men with tails and mothers birthing without borders – that is significant. Far more remarkable is the fact that Ireland has been frozen in time, as it were: the reference to 'ancient times' has been removed and replaced by a present continuous. More remarkable still are the cross-references to 'wondrous and monstrous things' in different parts of the world.

It could be argued that by freezing Ireland in time Fra Mauro, in a sense, does to it what Aeneas Silvius Piccolomini also tries to do: lock it 'out of this world'. The rest of the world, the real world, so to speak, is where things happen (Africa separating from India, for example), but Ireland is where nothing ever changes. It is a different world. By that same token, it could also be argued that Fra Mauro, like Aeneas, is suggesting that Ireland, all told, is not really worth remembering; and that point has indeed been made. The references to Solinus, Pomponius Mela and others, it has been argued, do not so much serve the purpose of lending weight to what Fra Mauro says, as they do of demonstrating that he has 'scant interest in such fanciful stories' and wishes to lay the blame for them on others.[12] If you believe *that* – he may be warning his viewers – you will believe just anything! I myself do not have time to waste on such monstrous things. I need to hurry on to more important matters. And you, my viewers, need to do the same. Sail down north and all you will find are the same old wonders – not to mention the 'bad men' of Iceland, or the 'fickle and fierce and cruel' Scots – but sail up south and whole new worlds will open up to you![13]

It is clear, however, that Fra Mauro had more than a passing interest in marvels himself, and knowing that his viewers also did, we may plausibly surmise that what he actually means to say is quite the opposite: *do* tarry! And do *not* dismiss Ireland, before you check Andaman (off the coast of India). There you will find that it says:

> the island of Andaman on the route with Taprobane to the east and the mainland to the west and between Paigu to the south and the north. Which island with its Andamaners has a circumference of about 500 miles and is inhabited by people who are idolatrous and sortilegers and cruel, and by many it is said that there is a lake on this island where, putting iron, it becomes gold, and this I say for the satisfaction of many witnesses.[14]

Witnesses, in other words, are to be contented. What they say, regarding marvels, should be believed, all the more so as Solinus, Pomponius Mela,

St Augustine, Albertus Magnus, St Thomas Aquinas and Pliny say so too. And that Ireland is there to prove it. What is for sure is that, on Fra Mauro's map, Ireland is a beacon of stability in a world where all is change, a place where, amid much doubt, one still 'can believe'. Or put otherwise, a place where, for one's good luck, the need-to-know-everything principle can be set aside, at least temporarily, where the truth *about* can still play second fiddle to the truth *for*. One may wish to speculate about its name – Fra Mauro calls it 'Hibernia', whereas 'some call it Ireland [*Irlandia*]' – and about the places of interest it may contain – 'purgatorio de san Patrizio' [St. Patrick's Purgatory], 'Garaforda' [Waterford?], 'Roa' [Ross?], '... orte' [?], 'Sancta I...' [Holy I[sland]?]. But all told one could take solace in the fact that there, and on the nearby islands, 'one finds many marvellous things', so that 'these islands of Hibernia are called fortunate'.[15] Admittedly, Fra Mauro does not say whether he himself would call Ireland fortunate, but neither does he say that he would not. He would certainly not have been the first Italian to call it thus.

Ever since antiquity, the Fortunate Isles had been thought to be the dwelling place of the blessed, an earthly paradise in the far western reaches of the world, half way between the herebelow and the hereafter, where the sun goes down in order to come up again. In the course of time these Isles had come to be identified, in people's minds, with what to us are the Canaries (from where, as we know, Ptolemy measured out his meridians), but they had also been believed to be located further north, roundabout where Ireland lies, and so it happened that there were those who still confused the Fortunate Isles with Ireland. Another reason for the confusion may have been the popularity of the legend of St Brendan, as told in the tenth-century *Navigatio Sancti Brendani* [*Voyage of St Brendan*]. According to the legend, in the sixth century St Brendan had set sail from Ireland with a group of fellow monks in search of the Promised Land of Saints, and after criss-crossing the seas to the west of Ireland for seven years, encountering many adventures on the way, he had discovered that fortunate isle in the middle of the ocean, beyond an everlasting mist. Originally written down in Latin, by an Irish monk living in Germany, the legend was soon to capture the imagination of other parts of Europe, in particular of Italy, and more especially of Venice.[16]

During the fourteenth and fifteenth centuries, the work was translated into Italian no fewer than five times, and in the oldest translation St Brendan is actually said to have been born in Venice, with many details added to the story to reflect the contemporary concerns of the Venetians. Perhaps that is what Fra Mauro had in mind when he called 'these islands of Hibernia' fortunate. But he may also have been thinking of what it said in another well-known work which spoke of Ireland's marvels, the *Topographia hibernica* [*Topography of Ireland*] by Gerald of Wales. Ireland, as we know, was already identified with marvels in the writings of the ancients, but it was Gerald of Wales who made the two synonymous, turning Ireland into marvel by antonomasia, and it was also he who was the first to hint that the Irish were fortunate indeed to be living in such a marvellous place. And what in Gerald was just a hint became, thanks to his first Italian reader – or if not his first, certainly his most famous and influential one: Francesco Petrarca – an established fact. Thanks to Petrarch, the Irish became a 'happy people' living in an earthly paradise that was at all costs to be visited.

* * *

To modern readers of the *Topographia* it may come as a surprise to find that Gerald's Irish are so fortunate. Nowadays we tend on the whole to view the work as an apology of colonialism. Gerald of Wales, otherwise known as Gerald de Barri or Giraldus Cambrensis (1146?–1223), was descended through his mother Angharad from Nest, the mistress of King Henry I (1068?–1100–35) and wife of Gerald of Windsor, Castellan of Pembroke in Wales, after whom the Fitzgeralds are named. He was an ecclesiastic, who spent thirteen years as a student in Paris, before returning to his native Wales and later retiring to Lincoln. In the course of his life he paid several visits to Ireland, in order to give assistance to Ireland's conqueror, Henry II, and to his son Lord John, or to visit his own relatives. His *Topographia*, which is written in Latin, was begun in the mid 1180s and almost immediately became a great success. By the time of his death, when it was still in the process of being revised and enlarged, it had well and truly established itself, in Britain and beyond, as the standard reference work on Ireland, a reputation it was to enjoy for many centuries to come. It is dedicated to Henry II, whom Gerald hails as 'our western Alexander', after Alexander the Great, the ancient conqueror of the world, and in essence it aims to show that Ireland is a land of plenty and of marvels, whose inhabitants are unworthy of the bounty God has bestowed on them and therefore need a civilized and Christian hand to guide them. So unappreciative are the Irish of that 'gift from God, on this earth incomparable', that Gerald has no qualms in claiming that, but for the conquerors, it would have remained unknown and, but for him, unsung:

> For just as the marvels of the East have through the work of certain authors come to the light of public notice, so the marvels of the West which, so far, have remained hidden away and almost unknown, may eventually find in me one to make them known even in these later days.

That, according to Gerald, is Ireland's real good fortune: to have been discovered by him![17]

The work is divided into three parts. The first, as though to make explicit the disconnect between the country and its inhabitants, deals exclusively with the land. The land, it says, is 'fruitful and rich in its fertile soil and plentiful harvests. Crops abound in the fields, flocks on the mountains, and wild animals in the woods'. Although it may suffer, more than any other country, 'from storms of wind and rain', it is 'rich in pastures and meadows, honey and milk, and wine, but not vineyards'. It has an abundance of wild animals, but no harmful ones, which makes 'some indulge in the pleasant conjecture that Saint Patrick and other saints of the land purged the island of all harmful animals'. It produces 'none but the best breed of falcons', and 'the air is so healthy that there is no disease-bearing cloud, or pestilential vapour, or corrupting breeze'. 'This is the most temperate of all countries. Cancer does not here drive you to take shade from its burning heat; nor does the cold of Capricorn send you rushing to the fire'.[18]

The second part of the work is dedicated to Ireland's miracles and wonders, which Gerald divides into the miracles wrought by nature and the wondrous deeds of saints. He talks, for instance, of the two islands in a lake in Munster, in one of which no-one dies, and in the other no animal of the female sex can go; or of the island in Connaught where corpses exposed in the open do not putrefy. What also makes Ireland special are its miraculous wells, such as the one in Munster which immediately turns the hair of anyone who washes in it to grey; and its lakes, like the one in

Ulster which had overflowed from a well when a woman, after drawing water, had forgotten to cover it again, and within an hour had drowned all the people of the area and their flocks. In addition, it has incredible animals or animal-like creatures: a wolf that talked with a priest, a woman with a beard and a mane on her back, a cow that was partly a stag, a goat that had intercourse with a woman, a lion that loved a woman. But above all it can boast of the miraculous works of its saints: St Kevin, St Colman, St Michael, St Nannan, St Yvor, St Finian, St Maidoch, St Finbar, St Patrick, St Luthernus, St Fechin, and, more than any other, 'the glorious Brigid'.[19]

Wonder-working though the Irish saints may be, Gerald is nevertheless keen to point out that their saintliness is in certain ways defective, as though he meant to question the legitimacy of their divine calling:

> This seems to me a thing to be noticed that just as the men of this country are during this mortal life more prone to anger and revenge than any other race, so in eternal death the saints of this land that have been elevated by their merits are more vindictive than the saints of any other region.[20]

Instead he vests that legitimacy in himself, by showing that it is in fact through him and his own work that God speaks. The most distinctive feature of the *Topographia* is Gerald's attempt to present anything marvellous happening in Ireland as a sign from God, which, for the benefit of all humans, it befalls him to interpret. There is thus a strong didactic and moralistic tone to the work.

One of the more famous marvels Gerald describes is the miracle of the so-called barnacle geese, that 'first appear as excrescences on fir-logs carried down upon the waters', then 'hang by their beaks from what seems like seaweed clinging to the log, while their bodies, to allow for their more unimpeded development, are enclosed in shells', and finally 'having put on a stout covering of feathers, either slip into the water, or take themselves in flight to the freedom of the air'. However, the real wonder of these birds, according to Gerald, is that they testify to the wonder of the Incarnation, and so he uses them as a pretext to launch into a violent tirade against the Jews:

> Pause, unhappy Jew! Pause – even if it be late. You hesitate to deny the first generation of man, from the slime of the earth without the co-operation of either man or

> woman; or the second, from man without the co-operation of woman – this, because of your veneration of the Law. The third only, that achieved by the co-operation of man and woman, because it is usual, you, with your hard neck, approve of and affirm. But the fourth generation, in which alone is salvation, that is from a woman without the co-operation of a man, you cannot, in your obstinate will, abide – and to your own destruction. Blush! wretch. Blush! At least consider the evidence of nature. She daily produces and brings forth new creatures without the co-operation of any male or female for our instruction and in confirmation of the Faith.[21]

Nowadays, of course, evidence of the sort would be dismissed out of hand, but in Gerald's own day, it would appear, the readers could not get enough of it. The many revisions which the work underwent over the years consisted primarily of the addition of ever more such moralizings.

To set the seal on his mission, in the third part of the work Gerald 'finally treats of man himself, as being the most worthy subject of our investigation, and on whose account we have treated of the other things'. But in saying so he is of course being disingenuous. 'Most worthy subject of investigation' though he may be, the type of man Gerald encounters in Ireland – or better, perhaps, deems to have encountered, or better still, wills to have encountered – has precious little of real worth about him, for as Gerald sees it, the people of Ireland

> is a barbarous people, literally barbarous. Judged according to modern ideas, they are uncultivated, not only in the external appearance of their dress, but also in their flowing hair and beards. All their habits are the habits of barbarians. Since conventions are formed from living together in society, and since they are so removed in these distant parts from the ordinary world of men, as if they were in another world altogether and consequently cut off from well-behaved and law-abiding people, they know only of the barbarous habits in which they were born and brought up, and embrace them as another nature. Their natural qualities are excellent. But almost everything acquired is deplorable.[22]

True, Gerald does recognize that the Irish have certain redeeming features. He notes, for instance, that they are skilled beyond compare at playing musical instruments and that their clergy are not without virtues. Nevertheless he finds them to be universally devoid of those features that are the hallmark of a civilized people: in the way they dress, the way they eat, the way they fight, the way they work (or do not), in their social, political and religious customs (or lack thereof), in their sexual mores, and in their moral attributes (or deficiencies). And their one redeeming feature, that is, their *potential* ('their natural qualities are excellent'), is in fact no more than an alibi for the invaders to march in. On their own the natives are incapable of improving themselves, they need the help of their betters.

Renaissance humanists would fully have agreed with such a view. Admirers of Rome that they were, they believed wholeheartedly in the civilizing mission of conquerors and empire-builders. As they saw it, that was what had made the grandeur of Rome. However humanists would have had no time for works like Gerald's – or at least that is what we would say nowadays. Is it not the case, indeed, that it was the humanists who invented the idea of the Middles Ages, that is, of the Dark Ages between the time of Rome and the time of Rome born-again? And is it not so that they therefore despised anything medieval as barbaric? Yet as it turns out, Gerald's work actually did meet with their approval, and it was read by them with great attention. And all thanks to Petrarch (1304–74), the man who is famous above all today as the poet who pined, in his sonnets, after his unattainable beloved Laura and relentlessly tortured his soul with doubts about the purity of that love.[23]

* * *

If anything did, it was Petrarch's indefatigable inquisitiveness and restlessness, and his unbounded fascination with all things Roman, which ushered in the Renaissance. Restlessness was instilled in him at an early age. Born in Arezzo, some forty miles to the south east of Florence, to a Florentine political exile, when he was still a young boy he moved with his father to Avignon, in southern France, which was then the seat of the papacy and thus a busy cosmopolitan centre, to which many people travelled from all parts of Christendom. Avignon was to be his home until he reached the age of fifty, but he paid visits to Italy on numerous occasions and on many others travelled to parts of Switzerland, Germany, northern France and the Low Countries, in eager search, most of the time, of literary treasures from the ancient world. And his return to Italy, in 1353, did not mark the end of his wanderings. Before finally putting a cap on his restlessness, at Arquà, in the Euganean Hills not far from Venice, where he spent the last six years of his life, he always kept on the move, as he commuted from court to court, in the northern part of the country, following the trail of princely patronage.

Travelling, for Petrarch, was not just a way of life; it was also a form of education, both literally and metaphorically speaking. Human beings were born to discover, that is, to seek out unknown lands, but also to push back the frontiers of knowledge, and that, in Petrarch's own words, implied the complicity of mind and foot. As he wrote in a letter to a friend, 'anyone wandering along the frontiers of the ancients, which are difficult of access but pleasant once you reach them, must often repress doubts of the mind underfoot'. The frontiers of our world, in other words, are both physical and mental. They are not simply places, but they are places we imagine, because we have been taught to do so, and the unknown we go in search of (say, Thule) must always be known to be unknown before we can set off. That is why any travel through the regions of the world must be accompanied by travel through the pages of a book. Machiavelli called this 'entering into the ancient courts of ancient men'. Petrarch called it 'talking with books', and he devoted years of his life to wandering from library to library in order to unearth new or better books to talk with. That is what really made the Renaissance: the unearthing of hidden literary treasures from the past, and the desire to make the world conform to what these works revealed.[24]

For that to pass it was of course necessary to spend much time secluded with one's books, rather than out about in the world, and humanists would always be torn between the call of *otium* (time off, as it were, withdrawal from the busy-ness of life) and the call of *negotium* (time on, participation in that busy-ness). For a good while after Petrarch, *negotium* would gain the upper hand. Humanists would make their peace with courts and cities, giving rise to what to us is known as civic humanism: humanism of and for the cities, learning applied to the resolution of the practical problems of daily life in society. Petrarch's, though, was more a humanism of the ivory tower. Personally he disliked cities. He considered them dens of iniquity and the nemesis of poetry. Instead he much preferred solitary places, as far as possible from the madding crowd, and woods were his place of predilection. 'The woods', he wrote in one of his Latin poems, 'are a joy to the muses, the city is the enemy of poets'. Such indeed was his love of woods [*silvae*] that his friends nicknamed him Woody [*Silvanus*], 'for owing to the hatred of cities and the love of woods which were bred in me from an early age, many of our friends, in their conversations, would call me Woody rather than Francis'. Only in woods was it possible to find the freedom [*libertas*] that allowed the human spirit to flourish. And that, as it turns out, was the promise of Ireland – or so at least said 'one Gerald, a courtier of King Henry II of England', in his 'booklet *On the Marvels of Ireland*'.[25]

Petrarch had not really gone in search of Ireland, or of Gerald. Indeed he claimed to know the place all too well, even suggesting that it was boringly predictable. What really interested him was Thule – the great Known Unknown – and it was on Thule, he wrote in a letter to a friend, one should focus all one's *Wanderlust*:

> Since, in truth, Britain and Ireland [*Hiberne*] and all the Orkneys in the western ocean towards the north, just as the Fortunate Isles in the same ocean towards the south, are known to us almost as well as Italy and France, partly through what we have seen and partly through the many accounts of travellers, we have begun to look around, and to wonder, and to investigate somewhat more carefully whether anywhere in the ocean there rises that island – Thule – of which all have sung in their writings.

In order to satisfy that *Wanderlust* – Petrarch goes on to explain in the letter – he had had 'a not pointless conversation' on the subject with someone called Richard, who had been Chancellor to the king of England – i.e. with Richard de Bury (1287–1345), a famous bibliophile and author of a work entitled *Philobiblon* [*The Love of Books*], whom he had met in Avignon in the 1330s – and Richard had assured him that there was much information on the subject to be found in the books which he owned at home and which he promised to let him have as soon as he returned to England. Unfortunately, despite many reminders, the promise was never kept. Such is the reliability of 'British friendship'![26]

In due course, however, there came into his hands

> a booklet *On the Marvels of Ireland*, written by one Gerald, a courtier of King Henry II of England, which, although it is threadbare in its contents, is nevertheless held together through no mean skill in the use of words; and what made it worthy of not being banished from my library altogether was a small section which contained a doubt about that very island, similar to ours and well worked through. Such similarity of mind therefore commended the author of the entire work to me. There he quotes

> the views of a number of authors according to which Thule is the last of the islands of the ocean that are close to Britain or between *Arthon* and the setting sun, where at the summer solstice there is no night, and at the winter solstice on the other hand no day, and beyond which lies the lazy and solid sea. He quotes Solinus and Isidore as witnesses of this, and yet he nevertheless asserts that that island is unknown to the west and that there is no island of such a nature or name over there. Therefore, following his conjecture, he reckons it to be either a famous but fabled [*fabulosa*] island, or to be cut off from other islands by endless space and thus to be looked for nowhere else but in the furthest recesses of the northern ocean. He cites Orosius as agreeing with this view. He could also have cited Claudian, where he speaks of 'Thule doomed by the hyperborean star'. But this aside, Giraldus makes his case adequately, arguing more or less in this way. You should examine the witnesses he quotes, to see what they say and to find out how trustworthy he is.[27]

Though he appears rather dismissive of the work as a whole, and though it may not have brought him any closer to finding out about Thule, Petrarch did in fact read the 'entire work', and not just the 'small section' on Thule which *On the Marvels of Ireland* does indeed contain. We can be sure he did because the copy of the work which he owned is today preserved in the French National Library in Paris and can be seen to contain many of the characteristic notes Petrarch is known to have inscribed in the books with which he liked to talk: flower-like asterisks, markers in the shape of a hand with outstretched index, short-hand glosses.

Such notes of course are not always easy to interpret. When Petrarch highlights something, without putting his reasons for doing so into words, he could just as easily be signalling his approval as his disapproval. Nevertheless a clear pattern does appear to emerge from all the marginalia. Unsurprisingly Petrarch takes the side of *imperium*. Where Gerald praises Henry II for being a new Alexander, Petrarch notes in the margin: 'Great praise of king'. And where Gerald eulogizes Caesar, Petrarch, who himself was to write a *Life of Caesar*, highlights the passage with the gloss: 'G. Cesar'. However, he is not interested in the many derogatory remarks Gerald makes about the Irish, nor in their alleged lack of faith, and neither in their many saints. Ireland's miracles do not attract his attention as such, but only when he can note parallels with miracles in Italy. For instance, where Gerald speaks of the well in Ulster, similar to one in Norway, which changes wood to stone, Petrarch remarks: 'the same is said about the river

Elsa in Tuscany'. Predictably perhaps, since he was an inveterate moralizer, he is also interested in the lessons Gerald draws from miracles. So for example he highlights, with two asterisks, where normally he would use just one, a passage in which Gerald tells of how a half-tamed falcon in Kildare came to trust human beings too much and so was eventually killed by a 'rustic', concluding that what this proves is that 'we should not be too confident that life will be forever enjoyable and cherished'.[28]

Petrarch is also interested in Ireland's missing fauna: certain fish, goats and above all bees, whose absence, or otherwise, merit no fewer than five glosses. Absent mice, or mice which should be absent, also get a special mention. With the word 'mus' ['mouse'] he draws attention to the island of Aran, west of Connaught, where 'mice are neither born nor can live if they are brought there. But should they happen to be brought there, they will at once make for the nearby sea and throw themselves into it. And if they are prevented from doing so, they will die on the spot'. And he singles out for particular approval the passage where Gerald notes: 'according to what Bede says only two types of harmful animals does the island have, namely wolves and foxes. We would add this third and extremely harmful one: mice'. 'I agree', writes Petrarch in the margin; and to ensure his approval would not go unheeded, just above he adds a second warning: 'mice'. What caught his fancy most, however, was quite a different missing wonder.[29]

When Petrarch was a child, in Avignon, he had gone to spend a summer in the Pyrenees, but the holiday had turned into a nightmare. Constant thunderstorms, with bolts so loud that they sent shivers down his spine and lightning so merciless that it set off countless forest fires, had left him traumatized for life. Ever since, even just hearing of the ravages that lightning could wreak would make the fear come back and haunt him. Such was his phobia of thunderstorms that he committed it to paper, in an exchange of poems on the matter between himself and his friends Boccaccio and Checco Rossi da Meleto (both of whom we shall meet soon again), and confessed it to St Augustine in his spiritual autobiography, *Secretum*, while also admitting that, deep down, his love for Laura, symbolized by the laurel tree, was sustained by the knowledge that the laurel was never struck by lightning, that it was a charm, we might say, against the dread of *coups de foudre* (*coups de foudre*, thunderbolts, is what the French call love at first sight). But in any case, what was the shame of suffering from such fears? Did not greater men suffer from them too? 'If you were to reprimand me', Petrarch tells St Augustine

> for being too fearful of the clap of lightning, since I could not deny it (and it is indeed not the least of the reasons for my love of the laurel that that tree, they say, cannot be struck by lightning), I would answer that Augustus Caesar suffered from the same disease.[30]

With the fear of that disease ever on his mind, when Petrarch read in Gerald that 'there are never any earthquakes in Ireland: in one year you will hardly hear it thunder once. No thunderbolts cause terror here, no lightning ever strikes', he could not, apparently, believe his eyes. Did such a place really exist? If so, what was he waiting for? Why did he not go there right away? And to bolster his resolve, beneath the sentence, at the bottom of the page, with clearly and carefully drawn letters ensuring that the gloss would stand out from all the others and make the passage spottable at a glance and memorable for all times, he wrote, speaking to himself: 'Go there, Woody! What are you waiting for?'[31]

After that, the thought of an island blessed by the absence of thunderstorms was never to leave him. It is not true – he was to write, as if to convince himself – that lightning is a sign of God's ire and that God's love therefore rains down more on lands where there are no storms and less on lands where there are, for if that were the case, why, that would mean that God prefers Egypt, 'the seat of all impiety and of all evil superstitions', to the Pyrenees, 'where Christ is worshipped'. Neither is it true that frequent thunderstorms, such as occur in Rome for instance, are an indication that the air is pure and clean. But even if it were, 'I for my part', he added, 'would most gladly exchange that purity with the Egyptians themselves or with the Irish, in whose lands it thunders very rarely indeed and only lightly, according to the testimony of those who write about these places or come from there'. A born-again Roman willing to swap Rome for Ireland, and what is more admitting as much in a letter the subject of which is the greatness of Rome? That is certainly a turn-up for the books. And Petrarch no doubt realized that it was, and he must have thought that he was probably losing the run of himself, for no sooner had he made the admission, and as though to balance the peremptoriness with which he had enjoined upon himself, in the manuscript of the *Topographia*, to rush to Ireland, than he declared: 'but I return to Rome'. He meant it metaphorically of course – 'I return to the subject of Rome' – but it is one of those metaphors that speaks literally of what one has on one's mind.[32]

There was also another aspect of life in Ireland which left an indelible mark on Petrarch's mind. For Gerald the surest sign that the Irish were barbarians was their refusal to subscribe to the civilizing process. They had turned their back on civilization – that is, on city-living, on the life of citizens [*cives*] – and instead had stubbornly opted to remain 'backwoodsmen' [*silvestres*]:

> While mankind normally progresses from the woods to the fields, and from the fields to settlements and communities of citizens, this people, despising the labours of agriculture, having little desire for the wealth of towns, and scorning the rights and obligations of citizenship, has wanted neither to abandon nor to reject the life which it has been accustomed to lead in woods and pastures.[33]

What is more, the Irish positively took delight, as they should not have done, in *otium* and *libertas*:

> They do not devote their lives to the processing of flax or wool, or to any kind of merchandise or mechanical art. For given only to leisure [*otium*] and devoted only to laziness, they think that the greatest pleasure is not to work, and the greatest wealth is to enjoy liberty [*libertas*]. This people, then, is a barbarous people, literally barbarous.[34]

For Gerald, it is clear, *otium* and *libertas* had quite a different meaning than they did for Petrarch. Where for Petrarch *otium* meant the peace and quiet that allowed the scholar-cum-poet to focus on his calling, and *libertas* meant the release from worldly constraints, for Gerald they meant laziness and licence, the unwillingness to sustain the civilizing process and the refusal to submit to duly constituted authority, or in short, the stubborn determination to remain *silvestres*.

Petrarch should really have agreed with Gerald, for he too was a great believer in the civilizing process (of Rome) and the imperative of obedience to duly constituted (imperial) authority. But he did not. For how could *Silvanus* disapprove of *silvestres*? And how could he disown *otium* and *libertas*? Yet he actually went further than not to disapprove or disown. In his manuscript of the *Topographia* he duly noted the above passages with the glosses 'Sol.' and 'otium', and soon after reading Gerald's 'booklet', it would seem, he decided to make a telling addition to the work *On the Life of Solitude* [*De vita solitaria*] which he had been in the process of composing for many years. *De vita solitaria*, which has been called 'a universal atlas of solitude', is a treatise which, approvingly, surveys all the individuals and peoples who chose to withdraw from the world, and amongst these, although seemingly giving scant importance to the matter, Petrarch, quoting Gerald's exact words, made up his mind to include the Irish:

> I leave aside Thule and Ireland, of which one, on account of the many who have written about it, is very famous but unknown, while the other, in truth, is very well known. The inhabitants of the latter, who despise riches and politics, and moreover have no time for agriculture, live – I know for a fact – in pastures and in woods; they take delight in leisure [*otium*] and their greatest wealth is liberty [*libertas*]. I would call them a happy people, if I were not prevented from doing so by some other shameful habit of theirs (assuming of course it is true) and the wickedness of their customs.[35]

Clearly, Petrarch had read enough of Gerald, as well as other works which spoke of the 'wickedness' of the Irish (such, for instance, as Pomponius Mela's *Cosmography*, which he was in fact the one to rediscover), not to be able to dismiss reports of such wickedness altogether. But his love of leisure and of liberty made him unwilling to give complete credence to them ('assuming of course it is true') and consequently he found himself able to take the logical step which Gerald himself had baulked at doing: if Ireland was an earthly paradise, did it not have to be the case that the Irish were a 'happy people'? Fortunate indeed was Ireland! Petrarch knew it 'for a fact'. It is little wonder that he should have decided not to banish *On the Marvels of Ireland* from his library.

However, that is not the actual reason which he gives for not doing so. What really tipped the balance – or so Petrarch says – was the 'similarity of mind' he discovered between Gerald and himself. What impressed him most about Gerald was how he knew to talk with books. True, his booklet was 'threadbare in its contents', but at least he had been able to hold it 'together through no mean skill in the use of words'. Gerald was a good writer. What is more, he did not take what he read at face value. He 'doubted', 'reckoned', made 'conjectures', quoted 'witnesses', 'examined' them. His, in short, was the investigative method, the search for truth *about*, typical of modern science, which at the time of Petrarch and his followers, foremost among them Machiavelli, and despite the pull of truth *for*, was emerging as the basis of (western) critical thinking. Gerald, in other words, even though he issued from the Dark Ages, was a proto-humanist. What is more, a proto-humanist who liked to moralize, to care for the Self.

* * *

So much did Petrarch approve of Gerald's booklet that he not only kept it in his library but recommended it and lent it to that other great master of Renaissance humanists, his good friend Giovanni Boccaccio (1313–75), who also read it carefully and also borrowed from it, causing no little surprise, it must be said, among modern critics, one of whom wondered why on earth Boccaccio should have been interested in 'such a mediocre little work'. What is even more surprising, according to another critic, is that Boccaccio should actually have been interested in Ireland. His masterpiece, the *Decameron*, a collection of one hundred short stories, whose heroes and villains roam the world as far and wide as did the Italian merchants and bankers who had made the wealth of Italy and for whose entertainment the work had been written, has been hailed as a paean to the 'heroic quest of merchants', and to prove just how heroic those merchants were, that critic pointed out that Boccaccio includes *even* Ireland among the settings for his stories.[36]

In truth, Ireland hardly figures in the *Decameron* at all. There is mention of it in only one story (Story Eight, Day Two), where it serves as a refuge for what today we would call an asylum-seeker, fleeing France because of an accusation of rape levelled against him, falsely, by the king's daughter-in-law. The place is hardly described at all, and yet we must assume, since Boccaccio was as concerned about the verisimilitude of his stories as any modern writer would be, that the mere choice of Ireland as a refuge is a sign that it was known and seen already then as a land of the welcomes. Just as no modern (European) novelist, keen to entice his readers with the plausibility of his stories, would dream of sending a character fleeing persecution to look for refuge in, say, Myanmar or Uzbekistan, since neither country is likely to figure on the average reader's radar, and certainly not as an earthly paradise, so Boccaccio too is unlikely to have dispatched his persecuted Frenchman to a place of which his readers had never heard or which they associated with fear and terror.

Boccaccio may have heard about Ireland from personal contacts, for he himself was born into that very world of Italian merchants and bankers, a world which had produced the 'heroes' who acted as 'the power behind the whole Norman adventure in Ireland' and a good few of whom still maintained strong commercial links with the island. From the village of Certaldo, not far from Florence, Boccaccio spent his youth and early manhood in Naples, where his father had been sent on business by the Florentine bank for which he worked. The city of Naples was at the time the largest and possibly most cosmopolitan city in Italy, and as the capital of an independent kingdom over which there ruled a royal family of French origin, which had brought with it the customs and culture of what was still to a large extent regarded as the most civilized country in Europe, it afforded the young Boccaccio the opportunity of being witness to a brilliant court and social life, which left a strong mark on him and is reflected in many of his works. It was in Naples, while training as an apprentice merchant and studying canon law at the university, that he embarked on what was to become a very successful literary career. Later, after a brief spell in Ravenna and Forlì in the north of Italy, and always wielding his prolific pen, he settled in Florence, where for close on twenty years he worked for the city government, being sent on many occasions to other parts of Italy or abroad, on official business. The last years of his life he spent in retirement in his native Certaldo, enduring poverty, but comforted by friendship and by his love of reading and of writing.[37]

It was from that friendship and that love that the idea sprang of composing the work for which Boccaccio made use of Petrarch's copy of Gerald's *On the Marvels of Ireland*. As they uncovered the past, humanists began to be confused by the mass of new information that was emerging and soon they felt the need for tools which would help them to organize and classify it. The tools they devised, which today would have been data-bases, could take on many forms. There were poems, or encyclopaedias, or histories, or treatises, or collections of tales, and among the more popular such works there were Petrarch's *Triumphs* [*Trionfi*], a who's-who of famous people from the distant and not-so-distant past, participating in an allegorical pageant in which Chastity triumphs over Love, Death over Chastity, Fame over Death, Time over Fame, and Eternity over Time. There were also Boccaccio's *On Famous Women* [*De mulieribus claris*] or his *Genealogies of the Gods of the Pagans* [*Genealogiae deorum gentilium*], which aimed to provide readers with all the information they needed regarding the great women from antiquity or ancient mythology. And then there was another work by Boccaccio, entitled *Of Mountains, Woods, Springs, Lakes, Rivers, Pools or Ponds, and of the Various Names of the Sea* [*De montibus, silvis, fontibus, lacubus, fluminibus, stagnis seu paludibus et de diversis nominibus maris*], the idea for which was suggested to him by Petrarch.

The purpose of the work is explained by Boccaccio as follows:

> when the name of a mountain is taken for a river, when that of a pond is taken for a mountain, or when woods are taken for towns or countries, it is easy for history to make no sense. And notwithstanding the fact that they are sometimes taken correctly, even learned men very often get their interpretations wrong, when they surmise that a mountain or a river that is in the west is to be found beneath the rising sun. To avoid which, and to remove the errors of incorrect interpretations, it is my intention to write down the mountains, woods, springs, lakes, rivers, pools or ponds and seas, wherever they may flow or stand beneath the sun, and such as I found them in any author of repute, especially pagan authors. And so that it will not be necessary for those investigating these matters to do much unrolling [of the manuscript], I will describe everything in alphabetical order. What is more, since woods grow and springs and rivers flow from mountains, and out of them we can see that lakes, pools and ponds are born, I have thought it not out of order to begin by writing of mountains.

De montibus, in other words, is a reading aid, aimed at assisting readers of unfamiliar texts to make sense of unfamiliar names. So, for instance, it can help them to remember that the *Tamesis* [Thames] is a 'very famous British river' but the *Sequana* [Seine] a very famous French one, or that they are not to confuse Mount *Ararath* in Armenia, where Noah landed the Ark, with the *Arar* in Germany, which is a river.[38]

It is divided into seven sections (one for mountains, one for woods, one for springs, and so on), with each section arranged alphabetically, and it consists of just under 2,000 entries, varying in length from half a line to three pages, of a modern edition, with the longest being on the Nile.

About half of the entries are on rivers, a quarter on mountains, around one hundred each on seas, springs and lakes, fewer on pools, and fewer still on woods, and in all there are twenty-two entries dealing with Ireland: two ponds and marshes (*Langelinum* and *Ruturgium*), four mountains (*Blagnia*, *Childardus*, *Salanga* and *Slanius*), and sixteen rivers (*Avelimphius*, *Avenimorus*, *Banna*, *Berva*, *Boandus*, *Cled*, *Eoirus*, *Phinnus*, *Moadus*, *Modarnus*, *Samairus*, *Saverennus*, *Sinneus*, *Siurus*, *Slana* and *Slicheus*). Very few of the Irish entries are longer than a few words, they generally provide the barest of information (the location and, sometimes, the origin of the name), and hardly any are particularly noteworthy. The most informative is the one on the Shannon:

> *SINNEUS* is a river in Ireland [*Hibernia*] which sallies forth from a large lake dividing the regions of Connaught and Munster. And since it is the king of the rivers of the island, not content with one bed, it throws itself into two parts, one of which, flowing south, washes the city of *Chillesonensem* and embraces the city of Limerick, and, having with its course separated the two Munsters, enters into the Brendanican ocean; the other, in truth, seizing a northern course and separating Meath and the further parts of Ulster from Connaught, staggers into the Northern ocean.[39]

The most striking is the one on the Bann:

> *BANNA* is a river in Ireland flowing through the region of Ulster. Indeed, emerging from a very large lake of the same region, it rushes into the northern ocean. And in this river a fish was once found which had the shape of a salmon and was so big that it could not in any way be dragged to the edge all in one piece.[40]

Except in two cases (*Childardus* and *Cled*), all the information about Ireland is taken from Gerald's *Topographia*, but unless they had read the work themselves, Boccaccio's readers would not have known so, for not once does Boccaccio actually quote his sources. Neither could the readers, unless they were extremely diligent, have formed any real picture of Ireland in their mind from reading *De montibus*, since every entry in the work is a separate entity, with no index or cross-references to help piece the various bits together.

These are just some of what the modern editor of the work has called its 'scientific disasters', which to us make it seem unreadable. To the readers of the time, however, it seemed anything but. We know of at least sixty-six extant manuscripts of it, dating from the fourteenth to the sixteenth century (none of them, unfortunately, in Boccaccio's own hand, which makes it impossible to be sure that the 'scientific disasters' are really his and not a scribe's), and between 1473 and 1532 it was published no fewer than ten times. It was also translated into Italian, Polish and Castilian, although translation is possibly too grand a term to use, at least with regard to the Italian version. Most names have been Italianized simply in their endings, the *Avelimphius*, for example, turning into the *Avelimphio*, and often the translator does not seem to have understood what he was translating. So, for instance, the *Moadus* flowing through 'the Connaught region' ['per Connactiam regionem'] becomes the *Moado* that flows through the 'Concanicha region' ['per la regione concanicia'].[41]

It is doubtful whether any of this really did assist learned men no longer to get their interpretations wrong, as Boccaccio had hoped, or whether it did help Ireland no longer to appear to be 'out of this world', but it cannot *not* have had the effect of confirming Gerald's status as an author of repute. Boccaccio's sources, as he himself points out, were mainly 'pagan', and the bulk of the information in *De montibus* is derived from Pliny and Pomponius Mela. For the 'mediocre little work' *On the Marvels of Ireland* to be admitted into such company was an accolade indeed, which was bound to impress anyone determined enough to have a serious talk with books. And so it came about. No sooner had the ink dried on *De montibus* than Gerald was pressed into service once again, and possibly through the very same copy of the work which Petrarch had lent to Boccaccio. However, the person who pressed him into service, a disciple of Boccaccio's, was apparently a tad embarrassed at having done so, and he therefore concealed Gerald's identity under the guise of an Irish bishop; and for proper measure, he himself took cover beneath the wing of his master. His name was Domenico Silvestri and he came from the same neighbourhood of Florence in which Boccaccio lived.

* * *

Very little is known about Silvestri (1335?–1411?). He was one of the group of civic humanists that took its inspiration from Coluccio Salutati (1331–1406), the Chancellor of the Florentine Republic (i.e. the head of its civil service, so to speak), and his career and interests reflected theirs. Most civic humanists were notaries by training, and so too was Silvestri. Most served in the administration of their state, and so too did Silvestri, as a middling civil servant, we would say today. And most, because of their mastery of language and their ability, as a result, to be effective communicators, were thought to be particularly well suited to become ambassadors; Silvestri too was sent on frequent embassies by his city. But above all civic humanists, like all humanists, believed in spreading the word, and so they taught and wrote.

Spreading the word often meant continuing the work of their masters, and it is for that that Silvestri is primarily remembered nowadays. He is the first translator into the vernacular of Petrarch's *Invectives against a Doctor* [*Invective contra medicum*], the manifesto, so to speak, of Renaissance humanism, which ridicules the sterile knowledge of old-fashioned academics who, because they had studied philosophy in order to become physicians, as required by the traditional university curriculum, had the temerity to hold and style themselves both physicians and philosophers, and, even worse, to 'make books in the midst of urine'. More importantly, as far as we are concerned, Silvestri sought to complete Boccaccio's *De montibus* with what could be called an appendix *On Islands* [*De insulis*], which he felt were of much greater interest than mountains, ponds and woods, and to which he devoted the better part of his life, producing a work which, though incomplete, is longer and has far more substance than Boccaccio's.[42]

Is it not surprising – muses Silvestri, in the introduction to *De insulis* – that 'our famous poet and fellow citizen', even though, 'ferried by the elegance of his prose, he often ploughed through gulfs and seas', should have shown not the slightest interest in islands? For is it not true that islands 'surpass woods, lakes and ponds and rivers, both in number and in terms of remarkable things'? And is it not the case that 'there is more to read about that is worth relating and likely to move one's mind to admiration and to be a serious read, more sights that are a joy to behold on islands than there are in ponds, pools, lakes or woods'? More to the point, who would deny that there is more that we can learn from islands, so as to make our life a better one, since 'an island is a land surrounded on all sides by the sea, and according to the true theologians it signifies the soul made steadfast by virtues, that is assailed by many temptations but, unmovingly steadfast, is not for turning'? Irrepressibly holier-than-thou, following the lead of Petrarch, and always inclined, as we have said, to teach one a lesson, humanists saw it as their mission to be useful to others; and it is in that spirit that *De insulis* is written, albeit in the more utilitarian kind of way typical of the post-Petrarchan generation. *De insulis*, Silvestri says, is intended for 'the use of its readers', and it is therefore written in 'common and current words', which are appropriate not 'for people who grow pale in quiet and leisure [*otium*] but for people who are suited to business [*negotium*]'.[43]

As it turns out, not many people were to benefit from Silvestri's endeavours to better them, at least not directly. There is only one known manuscript of *De insulis*, and no early printed edition. The manuscript is an autograph, apparently written in old age, and incomplete. Now housed in the National Library in Turin (where it was badly damaged by fire in 1904), it had originally belonged to Silvestri's parish church, that is, to the church of Santo Spirito in Florence. But there must have been at least one other one in circulation, for extensive use was made of the information gathered by Silvestri for the compilation of the section on islands in a massive encyclopaedia – *The Source of Memorable Things in the Universe* [*Fons memorabilium universi*] – written by a contemporary of his, Domenico Bandini or di Bandino from Arezzo (d. 1418). The *Fons* itself survives in approximately twenty-five manuscripts, all copied before 1460 and most written at the papal curia or within curial circles, so Silvestri's labours were not to be totally in vain, not least because his work is far more interesting than Bandini's. What Bandini writes about islands, or certainly Ireland, amounts to little more than a selective plagiarism of *De insulis*. Silvestri, for all his limitations, is an original writer.[44]

What makes *De insulis* particularly interesting is not so much the information it contains – although that is interesting too – as the light it throws on the process of data accumulation and interpretation which humanists had to carry out in order to make sense, anew, of the world. *De insulis* is very much a talking book, in the way in which Fra Mauro's map is a talking map. Silvestri, like Fra Mauro, struggles to make sense of 'marvels' and of 'novelties', and he often engages his readers in conversation, in language-games, to prod them and to challenge them. And he does so more particularly in connection with Ireland, for it is in Ireland above all that he is struck by the fact that time will not stand as still as the worshippers of 'ancient days' would like:

> all things change with time, and nothing is firm, nothing stable, and the customs of men and regions are as if caught up in a whirlwind and constantly changing and shifting. Today indeed the Irish have other habits and other customs, especially in the coastal regions, for although the places of this island upon the first arrival of guests seem inhospitable and hard to get to know, nevertheless in the coastal places, above all, they seem more gentle, although in some of the mountainous regions they have not entirely given up the habits of ancient days.[45]

As we know, of course, for humanists not every ancient day was equal to every other. The ancient days of what was 'out of this world' were far less equal than the ancient days of Rome. However, for Silvestri, who appears to have fallen under the spell of Petrarch's 'Go there Woody!' (although it must be admitted that we have no proof whatsoever that he was ever lent Petrarch's copy of the *Topographia*), the ancient days of Ireland would have been more than equal to his charmed expectations.

Following his master to the letter, Silvestri wrote a work which, like *De montibus*, is in 'alphabetical order, so that it is easier for anyone wishing to do so to find one of the islands'.[46] The work runs to more than eight hundred entries, the shortest consisting of just a few words, the longest of fourteen pages, of the modern edition, and to be on the safe side – for Silvestri too is mostly an armchair traveller and can only go by what others tell him – the entries also cover a number of what the French call almost-islands [*presqu'îles*], that is, peninsulas, as well as some might-be-islands, that is, places of which Silvestri is not too sure what they really are. Ireland is one of the islands which captures his attention most. It is afforded no fewer than five separate entries, for a total of seven pages, and is only superseded by Sicily (twenty-five pages and three entries: *Sicania*, *Sicilia*, *Trinacria*), England (ten pages and two entries: *Anglia*, *Britannia*) and France's Isle de la Cité in Paris (eight pages), with Taprobane (five pages), and *Tiros* (the city of Tyre in the Lebanon, which used to be an island) and Crete following close behind (four pages each). Most of the information is derived from 'pagan' authors, with Pliny, Pomponius Mela and Solinus topping the list of Silvestri's sources, followed in turn by Livy, Ovid, Virgil and Isidore of Seville. A number of more recent authors are also used, most notably Gerald of Wales and Boccaccio, but strangely, whereas all the other sources are duly acknowledged, not once is Gerald ever named by name.

The other surprising feature of the work is the presentation of Sicily, England and Ireland under more than one heading. In the case of Sicily an aesthetic justification is put forward. For humanists, the fundamental rule of writing was the injunction of the Latin poet Horace (65–8 BC), in his *Art of Poetry*, that a work be both 'useful' [*utile*] and 'sweet' [*dulce*]. Silvestri subscribed to this whole-heartedly. His ideal reader is a reader who, besides making full 'use' of what is written, is 'delighted' with what is written; to be a 'delighted reader' means not to have to endure 'tedium'; 'tedium' is best avoided by shunning 'prolixity'; therefore 'about *Sicania*, which is also called *Sicilia* and *Trinacria*, we will say a few words under each of these names so that prolixity does not generate tedium'.[47] In the case of Ireland, however, it is confusion rather than aesthetics which apparently led to the division.

The first Irish entry – or better, the first entry which *we* can tell for sure relates to Ireland – is headed *Iberia* and it begins: 'this island and Ireland [*Irlanda*] are the same, according to some; others say that Iberia and Spain [*Hyspania*] are the same, but since Orosius names it as a peninsula, therefore I make mention of it too. Others say that *Yberia* and *Ibernia* are the same'. The next entry, which comes right after *Iberia* and is the longest, is headed *Ibernia*, and here Silvestri says: '*Ibernia*, a large island, is close to Great Britain, as Solinus writes, but in fact is closer to Scotland and, as Orosius writes, is inhabited by Scots'. Further on in the entry he wonders whether, 'because *Ibernia* is close to England' and 'because the English are in charge of *Ibernia*', Munster [*Momonia*] and England may not in some way have become confused. In the third entry, headed *Irlanda*, there is still greater confusion, even though it is only a couple of lines long and Silvestri enlists the support of Fazio degli Uberti, a renowned expert on world affairs (whom we shall be meeting again shortly): 'Ireland [*Irlanda*] is a part or a region of *Ybernia*, as Fazio seems to imply, and towards the north it is close to Scotland, which some maintain is an island'. Only with regard to the fourth entry does Silvestri feel on safer ground; his readers, though, may not necessarily feel the same. It is dedicated to what Silvestri knows for sure is only a small part of Ireland, called *Midia* [Meath], but he still goes ahead and deals with it separately, 'not because Meath is an island in itself but because it is the most noble and fertile part of *Ibernia*'. Come the fifth entry, however, he is once again unsure. The entry is headed *Scotia* and it deals almost entirely with what to us is known as Scotland, but according to some Scotland and Ireland are actually one and the same place: '*Scotia* is an island, as Fazio degli Uberti claims, with whom Isidore agrees [...] Some however say it is a promontory of *Ybernia*'.[48] Talking with books, it is clear, could be an unforgiving task. The answers which they gave left a lot to be desired. But was there any better?

To resolve any doubts he may have, Silvestri goes through very much the same stages of denial, so to speak, as Fra Mauro was to do. As is to be expected, his first port of call are books. 'These and other wonders', he interjects at one point, 'would hardly be believable, if it were not for the fact that authors relate most of them'. Yet he is all too aware that what authors relate, others may interfere with. Time, 'than which nothing is more mutable', is the authors' prime enemy, changing 'the conditions and customs of places and men' to such an extent, that what was 'written by ancient authors' may now be 'different or have ceased to exist altogether'. On occasions, however, he is inclined to blame, if not the authors themselves, those whom Fra Mauro would call 'librarians': 'I do not know why there is such disagreement between authors, unless it is caused by the error of scribes'. In cases like these, the only way of resolving disagreements or solving doubts is to call on persons of experience. Best of all is personal experience, and it is that which allows Silvestri to give credence, for example, to a miracle in Leinster, where a river is said to produce stones in the shape of the crucified Christ and of John the Baptist. 'There is no small proof of that', Silvestri writes, 'in what I saw and discovered, and it is supported by what many witnesses have said', namely the stream not far from the town of Poggibonsi, south of Florence, where stones are often found in the shape of male and female pudenda, 'that cause not only admiration but awe in those who see them'. Failing personal experience, the next best thing is to rely on the witness of 'worthy men of learning', which is what Silvestri does in connection with the barnacle geese. When he had first heard about these strange creatures, his instinctive reaction had been to imagine that they were birds which had fallen from trees because, 'as is our custom', their wings were heavy with the pitch hunters had laid down to catch them; and nothing would have convinced him to believe otherwise, had he not 'seen three worthy men, of great learning and dignified manners, who declared and affirmed to have seen trees called Hibernians [*ibernas*] and to have eaten birds named barnacles'.[49]

No amount of seeing or of science or of good manners, however, could account for the two Irish lakes which, if a piece of wood was placed in them, changed that bit of it that was underwater into stone and the bit that was in the bed of the lake into iron, but left the rest intact. Lost for words, all Silvestri could do, in a case like that, was to appeal 'to the supreme craftsman, maker of all things, who rules and shakes heaven and earth, and whom the hidden deities of all the underworld hold in fear', for to him 'nothing is impossible'.[50] Yet appealing to God, as Fra Mauro was to find out, was in a sense too easy. God did not really provide explanations, and he was not a witness one could argue with. But what if, by untold good fortune, that 'supreme craftsman', to help resolve one's problems, should happen to send along one of his chosen few? To Fra Mauro he was to send some Coptic bishops. To Silvestri, on the other hand, he sent an Irish one.

Already on 'the island of *Dovillant*, situated in the British sea in the province of Golland or Holland', Silvestri is saved, as he struggles with the size of the island, by the intervention of one 'Bishop Milo', who is able to confirm to him 'viva voce' that it measures twelve miles in diameter. The good bishop then comes to rescue him again in *Iberia*, but it is only halfway through *Ibernia* that we are properly introduced to him. Whether we are to infer from the oversight that Milo was sufficiently well known at the time not to need an introduction or whether it simply means that *De insulis*, in the form it has come down to us, is indeed a work in progress, it is not really possible to tell. In truth, Bishop Milo is something of a riddle. That he was a bishop from Ireland, of that we can be fairly sure, but more than that it is difficult to figure out. He is presented as 'some Iberian by the name of Milo Elevane, bishop of that city on the island', and it has been suggested that he might be Milo Sweetman (d.1380), who in 1360 travelled through Florence on his way to Rome, where he was to receive papal confirmation of his appointment as Bishop of Ossory, although the pope then changed his mind and appointed him to the see of Armagh instead, which had become vacant in November of that year, following the death of Archbishop Richard FitzRalph.[51]

The year 1360 just about corresponds with the time when Boccaccio was putting the finishing touches to *De montibus* and Silvestri could therefore be sure that his master would not be dealing with islands. If the story about Bishop Milo is true, what it could mean is that it was the passage through Florence of an Irish bishop which actually prompted Silvestri to start work on the sequel to Boccaccio's dictionary. To have benefited from the testimony of a 'worthy' man of God to make sense, in part at least, of what was possibly the toughest riddle for an island-watcher to resolve – were *Iberia*, *Ibernia*, *Irlanda* and *Scotia* one and the same place? – may have made the enterprise seem much less daunting. However, the story as it stands does not really make that much sense. In particular it seems strange that Milo should have been going to Rome to meet the pope, when the pope was possibly in Avignon. So perhaps we should conclude that Silvestri's Milo is not Milo Sweetman, but Milo Cory, who was Bishop of Clonmacnoise from 1390 to 1397, assuming of course that Bishop Cory did visit Florence. If he did, a visit by him would have coincided with the very years (1385–1406) when Silvestri is believed to have been most actively involved in the composition of *De insulis*, and what is more, the popes by that stage had returned to Rome (or some popes, at least, for there were now rival papacies, in Rome and in Avignon).[52]

What is most intriguing about Milo is not only his identity or the fact that, all told, he does not really help Silvestri resolve the riddle of Hibernia (any Irishman, presumably, should have been able to tell Silvestri that *Scotia* is not 'a promontory of *Ybernia*'). It is that Milo does not really provide him with any new information about Ireland, as though the information Silvestri already had was all he really wanted, or needed to know. The only items regarding Ireland in *De insulis* which are not lifted straight from the pages of a book are the observations that the river which Boccaccio called *Sineus* is actually called *Sinen* in the language of the Irish, just as the 'famous city' of *Armoneca* [Armagh?] in Ulster is called *Elvensirelz*, as well as the clarifications regarding Solinus's description of Ireland. The Irish custom, described by Solinus, of victors painting their faces with the blood of their victims had continued to exist 'in the wooded parts until today', Silvestri was told, while other points mentioned by Solinus were incorrect: Ireland was now 'rich in birds and bees' and Irish sailors had much improved their skills. Everything else Silvestri writes about Ireland is copied from a written source, and one source in particular, for there is only really one topic which interests him: the marvels of Ireland.[53]

Under *Irlanda*, it is true, quoting Fazio degli Uberti, he mentions that Ireland exports serge to Italy, where it is used for making 'bedspreads',[54] and in *Midia* he talks of the exceptional fertility of the land, but otherwise his mind is entirely focused on miracles, be it the barnacle geese of *Iberia*, or the well-like hole in the ground beside the monastery of the Blessed Patrick which leads to the underworld, the lakes of *Ibernia* that turn wooden sticks to stone and iron, the river in Leinster which produces stones in the shape of the crucified Christ and John the Baptist, the werewolves of Connaught, the lake in *Midia* in which a dragon appears every seventh year to portend pestilence and famine, or the absent bees of *Scotia*, the stones and earth of which, wherever they be placed in a hive, cause bees to swarm away, and many many more. Such is Silvestri's fascination with Ireland's miracles, that at one point he loses the run of himself, spending almost half the entry on *Midia* recounting the true miracles not of Ireland but of other parts of the world, with Ireland serving as the litmus test, as on Fra Mauro's map some fifty years later, for the world's miraculous. If Ireland has miracles, so too must the rest of the world. If Ireland amazes, so too can the rest of the world.

Nevertheless, at other times Silvestri expresses doubts and would rather pass the buck. It is not he – he would have us believe – who insists on speaking of miracles and marvels, it is his pen: miracles 'will not remain inside the pen'. He even has doubts as to whether he should be writing about such matters at all. Reflecting on the barnacle geese and *Iberia*'s other miracles, he comments: these things are 'amazing to say [*mirabile dictu*], more amazing to believe, but well worth remembering [*memoratu dignum*]'.[55] The 'but' clearly says it all: the 'amazing to say' sits uncomfortably with the 'worth remembering'. And it is that very same unease we will find in Fra Mauro. Fra Mauro too will start from the premise that Ireland and marvels are synonymous, he too will take the marvels of Ireland as proof that marvels are for real, wherever they may be found, and he too will be unsure as to whether to allow his pen free rein. But when all is said and done, Fra Mauro, unlike Silvestri, will come to the conclusion that, in case of serious doubt, it is preferable for the *mirabile dictu* to 'remain inside the pen'.

Aeneas Silvius Piccolomini, for his part, will go one step further. Following the principle he enunciates in *De Asia* according to which history and fables make uncomfortable bed-fellows – 'In history we look for what is true and serious. Trifles we look for in fables' – he will eschew the *mirabile dictu* altogether and focus entirely on the *memoratu dignum*; and since Ireland is *mirabile dictu* by antonomasia, he will leave out Ireland altogether. By definition it is impossible for Ireland to be 'worth remembering'. Interestingly, the miracle Aeneas will go in search of while in Scotland, only to find that it has been 'relegated even further', is in fact an Irish miracle. It is Gerald's signature miracle, so to speak, the miracle of the barnacle geese. Quite why Aeneas should have gone in search of it in Scotland is not too clear. He was not the first to look for it where it was not – a century or so earlier Fazio degli Uberti claimed to have located it in 'Great Britain' – but what is for sure is that Pius, unlike other writers, did not confuse *Scotia* with *Hibernia*.[56] Perhaps, knowing that the two were close enough, not least in people's minds, and realizing that as long as Ireland remained 'worth remembering' so too would the 'amazing to say', he wished to settle the matter once and for all, by bringing to it the full weight of the eye-witness testimony of the man 'of repute' he himself held himself to be. Since marvels always flee 'even further' than the place – than the island – where they are believed to be found, there is no point for anyone to 'go there' in search of them. They are bound to be 'out of this world'.

Notes

1. See Falchetta's introduction, and extensive bibliography, in MAURO (2006); also Cattaneo (2011), Edson (2007).
2. The texts of the captions are all taken from MAURO (2006) and referred to by page: caption number. The English translations are my own, adapted from those in MAURO (2006). 'Qui se paga el datio' – 'Questo çirolo qui de soto è molto

pericolo per naveganti' – 'In questo luogo habitano mali homeni e non sono christiani' (Mauro (2006), pp. 269: 411, 229: 230, 713: 2896).

3. 'Isola nobilissima la qual se dice volta mia più de 4000 et è divisa in quatro regni, ne la qual se trova oro assai, piper, ganfora, legno aloe, el qual se chiama galambach e ha odor più nobel sia al mondo e vendesse a peso d'oro. E qui nasce in arbori uno fruto chiamado durian, è de grandeça de una rasonevel anguria e ha el scorço verde e gropoloso come la pigna et ha dentro v^{c} fructi, chadauno de grandeça de una rasonevel pigna e chadauno de questi v^{c} fructi hano differente suavità de sapor, e sono dentro de color paonaço e sono molto calidi. Qui se trova elefanti assai, e hi homeni de questa isola sono de mior condicion cha queli de le altre isole vicine e sono formosi homeni, forti e çentil e boni astrologi ma idolatri. Sono anchora maçor de statura che queli che nasceno in india e li lor elefanti maçor de queli de le indie, e queli de le indie maçor de queli che nasce ne le mauritanie, e questo è per el suo optimo sito e bontà d'aiere' (Mauro (2006), p. 181: 23).

4. 'Isola sochotra posta tra el mar de persia e aden ma più inverso el mar rosso e posta per contra la boca del dito mar. Hi habitanti solevano esser christiani e sono nicromanti e con quel arte tuo' e vende el vento prospero a hi naveganti' (Mauro (2006), p. 263: 385).

5. 'Credo che qui molti se meraveierà perché io meto el nascimento del nilo, ma certo se si moverano cum raxion e vorano investigar quanto ho fato mi e cum quela diligentia che qui dir non posso, i vederano che io me movo a demostrar questo per evidentissime chiareçe ho habuto' (Mauro (2006), p. 263: 389).

6. 'Molte opinion e leture se trova che in le parte meridional l'aqua non circunda questo nostro habitabile e temperado çona, ma aldando molte testimoniançe in contrario e maxime queli i qual la maiestà del Re de portogallo à mandato cum le sue caravele a çerchar e veder ad ochio, i qual dice haver circuito le spiaçe de garbin più de 2000 mia oltra el streto de çibelter [...]. E i diti hano fato nuove carte de quel navegar e hano posto nomi nuovi a fiumere, colfi, cavi, porti, di qual ne ho habuto copia. Unde se'l vorà contradir a questi i qual hano visto ad ochio, maçormente se porà non assentir né creder a queli che hano lassato in scriptis quelo hi non vete mai ad ochio, ma cusì hano opinado esser. Anchora io ho parlato cum persona digna de fede, che afferma haver scorso cum una nave de india per rabia de fortuna de traversà per zorni 40 fuora del mar d'india oltra el cavo de soffala e de le insule verde e qui pur al garbin e al ponente e per lo arbitrar de i suo astrologi i qual son lor guida i scorse circa 2000 mia [...]. Dice ancora Pomponio Mela nel terço libro de la sua Cosmographia che uno havea nome Eudoxo, el qual scampando Lathmin, Re de alexandria, usì del colfo arabico e navegò quela parte austral e vene fin a gades ch'è al streto de çibelterra. Adoncha

sença alguna dibutation se può affermar che questa parte austral e de garbin sia navigabile e che quel mar indiano sia occeano e non stagnon, e cusì affermano tuti queli che navegano quel mar e che habitano quele insule' (Mauro (2006), pp. 211–13: 149).

7. 'La diversità de li interpreti comete assai error in exponer li nomi. Unde ancora mi non me persuado molto in dimostrar tuta la verità de i nomi perché non è possibile acordar tanta varietà de lengue le qual secondo el suo idioma varia e confonde li nomi, i qual convien siano cambiadi e corrupti e per le lengue e per longeça de tempo et error de i librarij' (Mauro (2006), p. 627: 2435).
8. 'Perché ad alguni par da nuovo che io parli de questa parte meridional, la qual quasi està incognita a li antichi, perhò io respondo che tuto questo desegno da sayto in suso io l'ò habuto da queli proprij che sono nasudi qui, che sono stà religiosi, i qual cum le suo man me hano desegnato tute queste provincie e citade e fiumi e monti cum li suo nomi, le qual tute cosse non le ho possudo meter cum el debito ordine per non esservi logo' (Mauro (2006), p. 201: 98).
9. 'Perhò a l'Eterno Dio lasso la mesura de la sua opera la qual lui solo intende a ponto' – 'Et in queste sono molte novità la qual se tase' – 'se l'è vero quel che ho audito perché io non ardisco affermarlo' – 'Questa nobel e magnifica citade a voler narar le sue zentilixie e bei costumi el pareria incredibile e perhò qui el riman in la pena' (Mauro (2006), pp. 697: 2828, 187: 31, 295: 560, 445: 1375).
10. 'Nota che anticamente in questa ixola de ibernia laqual oltra mode è fertilissima se dici eserli un'aqua in laquale chi li mette dentro legno quella parte che è in la terra con tempo diventa ferro e quella parte che è circondata dall'aqua [diventa pietra e quella parte che è sopra l'aqua] reman legno e questo è certissimo de veduta. Più è in una di queste ixolette piccole qui dintorno che tutti quelli che li nassi in ditta ixola nassi con la coda come d'un castrone e sempre che le loro donne voliono parturire vanno fora de ditta ixila e'lli partorissi secondo li loro antichi e consueti costumi' (quoted in Mauro (2006), p. 739: CB2). The words between square brackets would seem to have been omitted in Mauro (2006). They appear in Almagià (1944), tav. XV. 'Dress rehearsal': Mauro (2006), p. 48.
11. 'In questa insula de hibernia, la qual oltra modo è fertilissima, se dice esser una aqua ne la qual chi li mete legno quela parte che è ne la terra cum tempo deventa ferro e quela è circumdada da l'aqua deventa piera e quela è sopra aqua riman legno, e se questo se crede se pò ancor creder de lago de andaman, e queli che de queste cosse mirabile e de altre monstruose desiderano esser copiosi, leçano Julio Solin polistoro, Pomponio Mela, sancto Agustino, Alberto Magno, sancto Thomaso de Aquin nel libro contra i curiosi, item le Metaure d'Aristotile e Plinio de le meraveie del mondo e vederano che de mile cosse non ne ho dite una' (Mauro (2006), p. 579: 2212).

12. Mauro (2006), p. 578.
13. 'Scotia [...] la gente è liçiera e feroce e crudel' – 'In questo luogo habitano mali homeni e non sono christiani' (Mauro (2006), pp. 671: 2697, 713:2896).
14. 'Insula andaman in dromo de taprobana da levante e da ponente de terra ferma e in dromo de paigu ostro e tramontana. La qual isola cum i suo andamani volta mia circa 500 et è habità per çente idolatra e sorteri e crudeli, e per molti se dice esser uno lago in questa isola che metandoli ferro deventa oro, e questo io dico a satisfation del testimoniar de molti' (Mauro (2006), p. 227: 220).
15. 'Alguni nomena questa irlandia' – 'In queste ixole se trova molte cosse meraveiose' – 'Queste insule de hibernia son dite fortunate' (Mauro (2006), pp. 671: 2689, 581: 2223, 583, 2227). For the other captions see Mauro (2006), pp. 581: 2225, 583: 2226, 583: 2229, 583: 2230–1, 671: 2690.
16. On the *Navigatio* and its popularity in Italy, and on the Fortunate Isles, see Bartoli (1993), Boscolo (1996–7), Grignani (1975), Humboldt (1992).
17. Gerald of Wales (1982), pp. 124, 55, 57. 'Alexander noster occidentalis' – 'O Dei donum in terris incomparabile!' – 'Ut sicut orientalium regionum prodigia, diligenti auctorum opera, in publicae notitiae lucem dudum prodiere, sic et occidentalia, hactenus quidem abdita fere et incognita, nostro tandem labore his vel occiduis temporibus inveniant editorem' (Gerald of Wales (1964), pp. 189, 70, 74). On Gerald see Bartlett (1982), Morgan (1999), Rollo (1995). Also Hadfield (1997), p. 92 ('Giraldus Cambrensis [...] the main apologist for Henry II's invasion'), Rubiés (2000), pp. 43–4.
18. Gerald of Wales (1982), pp. 34–5, 50, 47, 53. 'Gleba praepingui, uberique frugum proventu, felix terra est et fecunda. Frugibus arva, pecore montes, nemorosa feris abundant' – 'Aeolicis namque flatibus, et pluvialibus inundationibus, prae aliis terris haec exuberat' – 'Pascuis et pratis, melle et lacte, vinis, non vineis, dives est insula' – 'Quidam vero, figmento satis favorabili, conjectant, sanctum Patricium, aliosque terrae sanctos, pestiferis cunctis insulam purgasse' – 'Falcones Hibernia praeter generosos non habet' – 'Aeris quoque clementia tanta est, ut nec nebula inficiens, nec spiritus his pestilens, nec aura corrumpens' – 'Terra terrarum haec omnium temperatissima. Non cancri calor exaestuans compellit ad umbras; non ad focos capricorni rigor urgenter invitat' (Gerald of Wales (1964), pp. 26–8, 62, 56, 67, 66).
19. Gerald of Wales (1982), p. 81. 'Gloriosa Brigida' (Gerald of Wales (1964), p. 120).
20. Gerald of Wales (1982), p. 91. 'Hoc autem mihi notabile videtur, quod sicut nationis istius homines hac in vita mortali prae aliis gentibus impatientes et praecipites sunt ad vindictam, sic et in morte vitali meritis jam excelsi, prae aliarum regionum sanctis, animi vindicis esse videntur' (Gerald of Wales (1964), p. 137).

21. GERALD OF WALES (1982), pp. 41–2. 'Ex lignis namque abietinis, per aequora devolutis, primo quasi gummi nascuntur. Dehinc tanquam ab alga ligno cohaerente, conchilibus testis ad liberiorem formationem inclusae, per rostra dependent; et sic quousque processu temporis, firmam plumarum vestituram indutae, vel in aquas decidunt, vel in aeris libertatem volatu se transferunt [...]. Resipisce, infelix Judaee, resipisce vel sero. Primam hominis generationem ex limo sine mare et femina, secundamque ex mare sine femina, ob legis venerationem diffiteri non audes. Tertiam solam, ex mare scilicet et femina, quia usualis est, dura cervice approbas et affirmas. Quartam vero, in que sola salus est, ex femina scilicet sine mare, obstinata malitia in propriam perniciem detestaris. Erubesce, miser, erubesce; et saltem ad naturam recurre. Quae ad argumenta fidei, ad instructionem nostram, nova quotidie animalia sine omni mare vel femina procreat et producit' (GERALD OF WALES (1964), pp. 47–9).
22. GERALD OF WALES (1982), pp. 92, 102–3. 'Competens enim ordo deposcit, ut [...] tandem hominem ipsum, tanquam digniorem operis partem, et cujus gratia de ceteris disseruimus, introducamus' – 'Gens igitur haec gens barbara, et vere barbara. Quia non tantum barbaro vestium ritu, verum etiam comis et barbis luxuriantibus, juxta modernas novitates, incultissima; et omnes eorum mores barbarissimi sunt. Sed cum a convictu mores formentur, quoniam a communi terrarum orbe in his extremitatibus, tanquam in orbe quodam altero, sunt tam remoti, et a modestis et morigeratis populis tam segregati, solam nimirum barbariem in qua et nati sunt et nutriti sapiunt et assuescunt, et tanquam alteram naturam amplectuntur. Quod igitur in his naturae, illud optimum: quicquid fere industriae, illud pessimum' (GERALD OF WALES (1964), pp. 138, 152–3).
23. On Petrarch see Billanovich (1996), Wilkins (1961). Here is an example of the sympathy shown by humanists, and Petrarch in particular, towards (Roman) empire-builders: '... and these are the means whereby divine providence created the Roman empire, so that, for sure, it could be the head of the world' ['ecce quibus effectibus divina providentia romanum imperium fabricata est, ut esset scilicet mundi caput'] (*Familiaris* 15.9: PETRARCH (1968), vol. 3, p. 159). See also Introduction, note 15.
24. 'Perambulanti veterum confinia, accessu quidem aspera sed amena cum perveneris, creber scrupulus ingenii pede calandus est' – 'comites libros cum quibus loqui sum solitus' (*Familiaris* 3.1: PETRARCH (1968), vol. 1, pp. 105–7).
25. 'Silva placet musis, urbs est inimica poetis' (*Epistolae Metricae*, II.3.43: quoted in PETRARCH (1977a), pp. 246–7) – 'Propter insitum ab ineunte etate urbis odium amoremque silvarum multi ex nostris in omni sermone sepius me Silvanum quam Franciscum vocant' (*Familiaris* 10.4: PETRARCH (1968), vol. 2, pp. 305–6). For more on Petrarch's love of *otium* and solitude see *Familiaris* 15.8 (PETRARCH (1968), vol. 3, p. 157). On *otium* in the Renaissance see Vickers (1990).

26. 'Cum vero Britanniam et Hybernen cuntasque Orchades in occidentali occeano ad aquilonem, inque eodem Fortunatas insulas ad austrum, partim visu partim assiduo commeantium testimonio non aliter pene quam ipsam Italiam aut Gallias nosceremus, circumspicere mirarique cepimus aliquantoque scrupulosius inquirere sicubi undis emergeret celebrata hec omnium literis insula' – 'sermo non otiosus' – 'amicitia britannica' (*Familiaris* 3.1: PETRARCH (1968), vol. 1, pp. 105–6). On Petrarch's encounter with Richard de Bury see Bouloux (2002), pp. 174, 268; P. Boitani, 'Petrarch and the "barbari Britanni"', in McLaughlin Panizza and Hainsworth (2007), pp. 17–18.
27. 'Post annos vero venit in manus meas libellus *De mirabilibus Hybernie*, a Giraldo quodam, aulico Henrici secundi regis Anglorum, licet tenui rerum filo, non rudi tamen verborum arte contextus; quem ne totum bibliothece nostre foribus excluderem, brevis quedam ipsius particula promeruit, ubi de hac eadem insula nostre similis et operosa dubitatio inserta erat; itaque similitudo una ingenii michi totius operis commendavit auctorem. Aliquot ibi scriptorum sententias attingit, quod insularum occeani, que circa Britanniam que ve inter Arthon et occasum sunt, extrema sit Thile; ubi estivo nulla nox, brumali contra solstitio nulla dies; ultra quam pigrum atque concretum iaceat mare. Ad hec Solinum et Ysidorum testes citat; ignotam tamen nichilominus occidenti insulam, nullamque huius vel nature vel nominis illic esse confirmat; ideoque, coniecturam sequens, aut famosam sed fabulosam insulam existimat, aut infinito ab aliis spatio secretam, nusquamque alibi quam in intimis borealis occeani secessibus requirendam; cuius sententie astipulatorem facit Orosium. Poterat et Claudianum, ubi 'yperboreo damnatam sidere Thilen' ait. Hoc tamen omisso, rem agit, et ad hunc fere modum ille disputat. Tu testes quos inducit, examina quam sibi constent, et quanta sit dictis fides intelliges' (*Familiaris* 3.1: PETRARCH (1968), vol. 1, pp. 106–7).
28. 'Magna Regis La[us]' – 'Si[mi]le dicit[ur] de elsa tuscor[um] fluuio' – 'Ex quo patet casum in secundis fore metuendum, et vitae diuturnae delectabili et dilectae parum esse confidendum' (see GERALD OF WALES (1964), pp. 189, 191, 86, 123; (1982), pp. 124, 125, 62, 83–4).
29. 'Mus enim nec nascitur [...], nec vivit invectus. Sed si forte allatus fuerit, statim cursu praepeti in proximum mare se praecipitat. Sin autem impediatur, [statim] emoritur' – 'Duo tantum animalium genera nociva insulam habere Beda describit; lupos scilicet et vulpes. Nos tertium hoc addimus, et nocentissimum' – 'assentior' – 'mures' (GERALD OF WALES (1964), pp. 84, 61; (1982), pp. 62, 49–50).
30. Petrarch recounts his fear of lightning in *Familiaris* 15.9 (PETRARCH (1968), vol. 3, pp. 157–63). 'Si michi igitur exprobrasses quod adversus fulminis fragorem timidior sim, quia id negare non possem (est enim hec michi non ultima causa lauri diligende quod arborem hanc non fulminari traditur), respondissem Augustum Caesarem eodem morbo laborasse' (PETRARCH (1972), p. 160).

31. 'Terraemotus hic nunquam: vix semel in anno tonitruum audies. Non hic tonitrua terrent, non fulmina feriunt' – 'vade illuc siluane, q[ui]d stas' (GERALD OF WALES (1964), p. 72; (1982), p. 55).
32. 'Sede[s] omnis impietatis omniumque malarum superstitionum' – 'ubi Cristus adoratur' – 'pro virili mea parte libentissime nobilitatem hanc permutatam velim cum ipsis Egiptiis aut Hibernis, apud quos perraro etiam et leniter tonare, et qui de his scribunt et qui inde veniunt testantur' – 'sed revertor Romam' (*Familiaris* 15.9: PETRARCH (1968), vol. 3, p. 161).
33. GERALD OF WALES (1982), pp. 101–2. 'Cum enim a silvis ad agros, ab agris ad villas, civiumque convictus, humani generis ordo processerit, gens haec, agriculturae labores aspernans, et civiles gazas parum affectans, civiumque jura multum detrectans, in silvis et pascuis vitam quam hactenus assueverat nec desuescere novit nec descire' (GERALD OF WALES (1964), p. 151).
34. GERALD OF WALES (1982), p. 102. 'Item non lino vel lanificio, non aliquo mercimoniorum genere, nec ulla mechanicarum artium specie vitam producunt. Solum enim otio dediti, solum desidiae dati, summas reputant delicias labore carere, summas divitias libertate gaudere. Gens igitur haec gens barbara, et vere barbara' (GERALD OF WALES (1964), p. 152).
35. 'Transeo Thilem et Hibernem, quarum altera scribentium varietate famosissima sed ignota est, altera vero notissima. Cuius gentem, opum rerumque civilium contemptricem, insuper et agriculture negligentem, pascua et silvas incolere compertum habeo, cui pro delitiis otium, pro summis opibus sit libertas. Felicem gentem dicerem, nisi alia, si modo vera est, infamia et morum malignitate retraherer' (PETRARCH (1977a), p. 238). On the date of *De vita solitaria* see Martellotti's introduction to PETRARCH [1977a]; also Kirkham and Maggi (2009), pp. 179–95 ('universal atlas', p. 181). For more on Petrarch's reading of Gerald of Wales see Haywood (2006).
36. On Petrarch's loan to Boccaccio of the 'mediocre operetta' see Pastore Stocchi (1963), p. 66. '... la *quête* eroica dei mercanti [...] la vastità dei campi e la ricchezza delle città [...] persino di Scozia e di Irlanda' (Branca (1994), p. 50).
37. 'The power ...': see O'Sullivan (1962), p. 101. See too Carus-Wilson (1967), Melis (1990). On Boccaccio see Branca (1976).
38. 'Dum montis nomen pro flumine, dum paludis pro monte, seu silve civitatis vel provincie loco sumitur, facile sensus hystorialis confunditur. Et esto aliquando proprie assumatur, existimatione etiam eruditi persepe falluntur, dum montem seu fluvium qui in occiduo est arbitrantur sub oriente sole consistere. Ad evitationem quorum et indebite existimationis amovendos errores, celebres inter auctores quoscunque et potissime gentiles montes, silvas, fontes, lacus, fluvios, stagna seu paludes et maria et quo sub celo effluant et consistant ut comperi

scribere mens est. Et ne multum talia scrutantibus evolvere necesse sit, per ordinem licterarum alphabeti describam singula. Sane, quoniam e montibus excrescere silvas et manare fontes et flumina a quibus lacus, paludes et stagna cernimus exoriri, de montibus primum scribendum non incongrue ratus sum' – 'TAMESIS celeberrimus Britannis est fluvius' (Boccaccio (1998), pp. 1827–8, 1975). On Petrarch's influence on Boccaccio, and more generally on *De montibus*, see Pastore Stocchi's introduction to Boccaccio (1998).

39. 'SINNEUS fluvius est Hibernie ex lacu maximo Connaciam Momoniamque regiones disterminante prorumpens. Et cum inter ceteros insule obtineat principatum, non uno contentus alveo in duas effunditur partes, quarum una versus austrum tendens Chillesonensem alluit civitatem et complectitur Limericum, distinctisque meatu duabus Momoniis Brendanicum ingreditur Occeanum; reliqua vero in arctoum iter arripiens Mediam et ulteriores Ultonie partes a Connacia segregans in Occeanum labitur Boreum' (Boccaccio (1998), p. 1973).

40. 'BANNA fluvius est Hibernie per Ultoniam effluens regionem. Nam exiens ex lacu eiusdem provincie permaximo in borealem Occeanum decurrit. In hoc enim olim piscis repertus est formam salmonis habens, tante magnitudinis ut nullo modo integer in margine trahi posset' (Boccaccio (1998), p. 1920).

41. Boccaccio (1510?), pp. XXXIIr, XLIIIr. On the quality and the early editions of the work see Pastore Stocchi's introduction to Boccaccio (1998).

42. On Silvestri see Cardini (1987), Milanesi (1993), Pecoraro's introduction to Silvestri (1955), Ricci (1950). Pouring scorn on would-be philosophers, in the *Invectives* Petrarch says: 'Why are books not made in the course of banquets? They are made in the midst of urine' ['cur non inter epulas libri fiant? Inter urinas fiunt']: Petrarch (1972), pp. 56–7. Silvestri's translation of the *Invectives* is published in Petrarch (1950).

43. 'Concivis noster et poeta preclarus [...] quot quandoque sinus et maria ornato suo navigio sulcaverit [...] cum silvas, lacus et paludes et flumina excedere videantur et in numero et in rebus magnificis [...] plura enim et relatu digna mentemque admiratione motura et lectu seria scituque iocunda gesta ac visa leguntur in insulis quam paludibus, stagnis, lacubus vel in silvis [...] ad usum legentium [...] popularibus et usitatis verbis et non quieti otioque pallentibus, sed negotiis convenientibus' – 'Insula quidem est terra undique mari circumsepta, et secundum veros theologos significat animan constantem virtutibus que crebris vexatur tentationibus, sed firma in constantia non volvitur' (Silvestri (1955), pp. 29, 31).

44. On Bandini see Hankey (1957), Hankey (1960).

45. 'Omnia mutantur tempore, et nihil solidum, nihil stabile et hominum regionumque mores turbinis more coluntur et mutuis varietatibus alternantur. Hodie

enim alios ritus, alios mores habent et maxime circa partes marittimas, nam licet huius loci insule primo hospitum adventu inhospiti atque asperi videantur conversatione, tamen in maritimis precipue locis mitiores apparent, licet in montanis quibusdam antiquitatis ritus non in totum demiserint' (SILVESTRI (1955), p. 128).

46. 'Boccaccius quidem imitatus elementorum ordinem sequar ut agilius sit aliqua [*recte?* aliquam] ex insulis invenire volentibus reperire' (SILVESTRI (1955), p. 33).

47. 'He has won every vote who has blended profit and pleasure, at once delighting and instructing the reader' ['omne tulit punctum qui miscuit utile dulci,/lectorem delectando pariterque monendo']: HORACE (1999), pp. 478–9, ll. 343–4. 'Lector delectatus' – 'Sicania que etiam Sicilia dicitur et Trinacria de qua sub quolibet horum nominum ne rei prolixitas tedium generaret aliqua tangemus' (SILVESTRI (1955): Sicilia, Sicania).

48. 'Hec insula et Irlanda idem sunt, secundum quosdam; alii dicunt Iberiam et Hyspaniam idem esse, sed cum Orosius peninsulam eam nominet ideo de ea mentionem feci. Alii dicunt Yberiam et Iberniam idem esse' – 'Ibernia magna insula est prope Britanniam magnam, ut Solinus scribit, multo vero vicinior Scotie a Scotis, ut Orosius scribit, habitata' – 'quia Ibernia est Anglie proxima [...] quia Angli presunt Ibernie' – 'Irlanda pars seu regio Ybernie est, ut Fatius videtur innuere et est versus septentrionem Scotie vicina quam aliqui asserunt insulam' – 'Midia non quia per se insula sit, sed quia pars Ibernie nobilior fertiliorque est inter insulas pono' – 'Scotia insula est, ut vult Fatius de Ubertis cum quo consonat Ysidorus [...] aliqui vero dicunt promuntorium esse Ybernie' (SILVESTRI (1955), pp. 126–7, 130, 138, 164, 206).

49. 'Hec et alia mira vix quidem credenda nisi autores plurima tradant' – 'hec et alia plura que in Ybernia tetigimus ab antiquis autoribus scripta reperio, sed quia tempore quo nichil alterabilius conditiones moresque locorum et hominum variantur, forsan horum plura aut aliter sunt, aut desierunt omnino' – 'Nescio tamen unde tanta sit inter autores discordia nisi scriptorum error eam faceret' – 'huius fidem rei facit non modicam quod vidi et repperi et multorum testimonium auditu verificat' – 'ut videntibus admirationem non solum sed stuporem incutiant' – 'nisi tres viderim dignitate moribus et scientiam [*recte?* scientia] venerandos asserentes affirmantesque arbores ibernas vocatas vidisse et aves bernaccas nomine comedisse' (SILVESTRI (1955), pp. 165, 170, 129, 127).

50. 'sed apud summum opificem omnium conditorem qui celum terramque regit et tremit et [*quem?*] omnium infernorum abdita numina perhorrescunt nil impossibile constat' (SILVESTRI (1955), p. 128).

51. 'Dovillant insula in Britannico mari sita in provincia Golandie sue Olandie [...] Milo episcopus [...] michi asseruit viva voce' – 'ab iberico quodam nomine Milo

Elevane insule, illius civitatis episcopo' (SILVESTRI (1955), pp. 97, 128). On Milo Sweetman see Smyth (1966).

52. On Milo Cory see Ware (1764), p. 172.
53. 'Lacus Regum, ex quo maximum flumen egreditur Sinen appellatum, seu secundum Boccaccium Sineus' – 'non longe ab Armoneca famose civitate Ultonie, que eorum lingua vocatur Elvensirelz' – 'in locis silvestribus [...] usque hodie' – 'hodie vero [...] avibus et apibus ferax' (SILVESTRI (1955), pp. 165, 129, 127).
54. 'operimenta lectorum' (SILVESTRI (1955), p. 138).
55. 'non calamo remanebit' – 'mirabile dictu, creditu mirabilius sed memoratu dignum' (SILVESTRI (1955), pp. 126–7).
56. 'Nugas in fabulis, in historia verum quaerimus & serium' (PIUS (1571?), p. 281). On Pius and the barnacle geese see above Chapter 1. '... ne la gran Bretagna,/ [...] fama è [...]/ch'arbor vi sono di tal maraviglia,/che fanno uccelli: e questo è il frutto loro' ['in great Britain fame has it that there are such wondrous trees that they produce birds; and that is their fruit']: UBERTI (1952), vol. 1, pp. 319–20: IV, xxiii, 1, 37–9.

CHAPTER THREE

The Other World
Imagining Ireland in the Shadow of Dante

Authors and works discussed in this chapter (in chronological order):

- DANTE ALIGHIERI (1265–1321), *The Divine Comedy* [*La divina commedia*]
- RANULF HIGDEN (d. 1363), *Polychronicon*
- FAZIO DEGLI UBERTI (1301?–67?), *Dittamondo*
- LEONARDO BRUNI (1370–1444), *Panegyric of the City of Florence* [*Laudatio florentinae urbis*]; *Funerary Oration for the Florentine Knight Nanni Strozzi* [*Oratio in funere Nannis Stroze equitis florentini*]
- CRISTOFORO LANDINO (1424–92), *The Commentary by the Florentine Cristoforo Landino upon the Comedy of the Florentine Poet Dante Alighieri* [*Comento di Christophoro Landino fiorentino sopra la Comedia di Danthe Alighieri poeta fiorentino*]
- FRANCESCO BERLINGHIERI (1440–1501), *The Geography of the Florentine Francesco Berlinghieri Produced in Terza Rima and the Tuscan Language [...] according to the Geography and the Maps Produced by Ptolemy* [*Geographia di Francesco Berlinghieri fiorentino in terza rima et lingua toscana distincta [...] secondo la Geographia et distinctione de le tavole di Ptolomeo*]
- MATTEO MARIA BOIARDO (1441–94), *Orlando Innamorato*

What was referred to in the previous chapter, flippantly perhaps, as 'cross-examining God' was a symptom of the characteristic Renaissance phenomenon of 'seeking to establish a meaning outside the sacred'.[1] Contrary to what was once held to be the case, the Renaissance was not a godless era, far from it,[2] but what did certainly make the age different from the preceding

'Dark Ages' was the attempt, spearheaded by born-again Romans, to provide secular answers to the big questions of life, answers, that is, which were demonstrable by human reasoning – by language-games – and which, insofar as it was possible or necessary, sidelined the divine and the sacred. Fra Mauro's solution to the question of the earthly paradise is a case in point. His caption shows him not to be entirely convinced of its existence:

> The Paradise of Delights not only has a spiritual meaning but is a place located on earth, says Augustine in his work on *Genesis*, and also in his book on *The City of God*, and that place is very far from human settlement and knowledge, situated in the eastern parts, according to the teachings of the holy doctor Bede.[3]

Yet to deny the existence of the earthly paradise would have been anathema, especially for a monk. So Fra Mauro includes it within the frame but places it outside the map proper, as if to say: the earthly paradise may well be of this world, but if it is, it is most likely in a place which is beyond the reach of human minds and thus – to echo Petrarch – of human feet. But did it really matter? Since only God Everlasting can understand the full measure of His work, why not leave the spiritual meanings to Him and let human knowledge focus on that alone which is located on earth? Or would that be tantamount to denying God?

For sure, going 'outside the sacred' was not a risk-free or an easy task, and for Italians – Tuscans and Florentines in particular – the task was made all the more difficult in that they had to cross-examine, to contend with, not only God but also Dante Alighieri (1265–1321). Dante was the geographer *par excellence*. The compelling vision of the world and the otherworld he had depicted with such mastery in his stunning *Divine Comedy* left no room for doubt. It contained and accounted for every part of God's creation (the earth, the heavens, hell, purgatory and paradise) as well as every one of its inhabitants, past, present and future. At the same time it warned of the perils and pointlessness of travelling – without a valid travel-pass, so to speak. In Dante's scheme of things anyone attempting to achieve what the Portuguese were to do with the help of Fra Mauro's map would have been consigned for evermore to the lowest depths of hell. Clearly if travelling and discovering were to have a guiltless future, Dante, while being acknowledged as the master he was, somehow had to be tamed and brought down to earth. As long as that did not happen, explorers would be doomed, Ulysses-like.

In Canto Twenty-Six of Dante's *Inferno*, among the evil counsellors and wrapped in a tongue of flame for all eternity, there marches Ulysses, the brain behind the Trojan horse. However he is not so much the hero of Homer's *Odyssey*, which Dante did not know, as the creation of imagination and of legend. Having brought about the fall of Troy, this Ulysses, rather than returning home to his faithful Penelope, chose to sail around the Mediterranean, drawn by the curiosity of seeing unknown lands, and when at last he reached the Pillars of Hercules (i.e. the Strait of Gibraltar), instead of turning back, as he should have done, since the Pillars marked the spot beyond which humans were not to venture, he urged himself and his companions on, in the name of virtue and of knowledge. 'Do not deny us the experience, following the sun, of the world without people', he harangued his companions, 'consider your seed: you were not made to live like brutes, but to follow virtue and knowledge'. Unable to resist, they thus set off on a 'mad flight', which took them through the Pillars and then ever further towards the 'left-hand side' until, beneath the 'other pole', they beheld in the distance a mountain so tall they had never seen anything like it. They rejoiced, but soon their joy turned to sorrow, for a storm arose from this 'new land' which struck their ship and dragged them to the bottom of the sea.[4]

That tall mountain was the mountain on top of which lay the Garden of Eden. In Dante's world, where otherwise the landmass was all concentrated in the northern hemisphere, it stood at the antipodes of Jerusalem. It had been formed by all the matter which had fled in horror before Lucifer, when he was expelled from heaven and hurled by God to the very centre of the earth, where he was now entrapped up to his waist in a sea of ice. The fleeing matter had left behind a gaping hole, shaped like a funnel, which stretched from beneath the surface of the earth to where Lucifer stood forever frozen in his evil. Made of ten concentric circles, where sinners earned the wages of their sins through everlasting torments, that hole was hell. The mountain in due course, when Christ had come, would be Mount Purgatory, on the seven terraces of which repentant sinners earned admission to paradise through penances of varying intensity and length, depending on the gravity of their trespasses. As for paradise, it was in heaven – or in the tenth heaven, to be precise. According to Dante's understanding (which was in fact derived from Ptolemy: Ptolemy the astronomer), around the earth, which stood motionless at the centre of the universe, there rotated nine heavens, one for each of the seven planets (in ascending order: the moon, Mercury, Venus, the sun, Mars, Jupiter, Saturn), one for the so-called fixed stars, and one which was the Prime Moved and imparted movement to the other heavens through its own perpetual motion, brought about by its desire to be fully and constantly in touch with God, who, if He was anywhere specific, was everywhere in the Empyrean, the infinite tenth heaven, which contained the other nine and the earth.

Only after Christ had come could any human be admitted to purgatory or heaven, so, however remarkable it may have been in itself, Ulysses's 'mad flight' never stood a chance of succeeding. There was only one way of reaching Mount Purgatory, the Garden of Eden and paradise, and that was Dante's way: down through hell, or in other words, through a (Christian) understanding, and rejection of sin. Purgatory – salvation – was not for the simply curious. In one respect, though, Ulysses had been right and hence deserving of Dante's admiration: the human seed was *not* the seed of brutes, it was of its nature drawn to 'mad flights' of discovery, and such flights were designed to pursue both knowledge *and* virtue. That, in essence, is what the *Comedy* is about. It is the account of a flight in pursuit of knowledge and virtue – the totality of knowledge, and virtue in its absolute perfection (Christian virtue). It is a travelogue which tells of a journey of discovery – through the otherworld – the aim of which is self-improvement and ultimately the discovery of truth and God. The work was to be enormously popular, and influential in many ways, but there was one lesson in particular which later generations drew from it and that was that travelling and discovery – and by extension geography – could not be self-sufficient. They had to have a moral goal; they pertained to the realm of ethics, just as much as they did to the realm of physics and metaphysics. Renaissance humanists in particular would take the lesson to heart, obsessively committed as they were to what was good *for* human beings.

One of the very first writers to give expression to this notion, in truly Dantesque fashion, was a contemporary of Petrarch's, who may or may not strictly speaking have been a humanist, but whose work enjoyed great popularity in the fifteenth and early sixteenth centuries and who came to be seen as an authority on what we might call world affairs. Nowadays the work is hardly known at all, except by a few specialists, and to a modern reader it is bound to seem dated and uninspiring, a clumsy attempt to out-Dante Dante. Perhaps readers of the time also found it wanting, poetically speaking, but it was its contents above all which made it invaluable. Domenico Silvestri quoted it in support of the view that *Irlanda* was 'a part or a region of *Ybernia*'. Aeneas Silvius Piccolomini may have had it in mind when he went to look for barnacle geese in Scotland. Fra Mauro referred to it in order to lend weight to the idea that Africa might be circumnavigable: 'Ethiopia [i.e. Africa] above, in the east is linked to the forests of Egypt', the work says, 'and towards the north it hears the Nile. In the south it is bathed by the Ocean and towards the west it extends so far that it touches Atalantes with its heels'. And in *Irlanda-Ybernia* itself it was cited as late as 1786, by Lord Charlemont, the first president of the Royal Dublin Society, who mentioned it in a paper on the 'Antiquity of Woollen Manufacture in Ireland' delivered to the Society's members to prove the estimation in which Irish cloth was held abroad: '*Irlanda*', says the work, 'among us is worthy of fame, for the noble serge it sends us'.[5]

The work is entitled *Dittamondo* [*World Show*: from *dittare*, to point out, to show, and *mondo*, the world]. It was begun in the 1340s, is written in the vernacular (Tuscan) and composed in *terza rima* (the rhyming scheme invented by Dante), it tells of a Dante-like journey of discovery undertaken under the guidance of a sage from antiquity (Solinus, in this case, rather than Dante's Virgil), it is even longer than the *Comedy*, though incomplete (consisting of 15,308 lines compared to the *Comedy*'s 14,083), and it survives today in no fewer than fifty-seven manuscripts (the majority Italian, with six dating from the fourteenth, forty-nine from the fifteenth, and one each from the sixteenth and eighteenth centuries) as well as two early printed editions (Vicenza 1474, Venice 1501). Its author is Fazio degli Uberti (1301?–67?), about whom little is known, except that he came from a family of Florentine exiles and that, like many an exile, his lot was to wander from city to city, hiring out his knowledge and his pen to any man of power who had need of it and could afford it. He was born in Pisa, it is thought, and at one time or another resided in Verona, Padua, Treviso and Bologna, but it was in Milan, at the court of the Visconti, the city's rulers, that he lived the longest and it is there that he mostly wrote his masterpiece.[6]

What we do know for sure about Fazio is that he counted among his ancestors another of Dante's heroic sinners, Farinata degli Uberti (d. 1264), the man who, through political miscalculation or bad luck, was in fact responsible for the family's exile. Described by Dante as 'magnanimous', Farinata, in Canto Ten of *Inferno*, is sentenced as a heretic to be enclosed for evermore into a sealed tomb. But the punishment does not frighten him. When Dante visits him (it is before the Day of Judgement and so the tomb, in a cruel twist of the punishment, is still unsealed), he shows by his language and his demeanour that he holds hell 'in great contempt'.[7] It is tempting to imagine that that is what prompted his descendant to write the *Dittamondo*. Dante, even though he belonged to the opposite faction, gives credit to Farinata for his (Florentine) patriotism, but still he condemns him to everlasting damnation. Fazio, out of gratitude to Dante, would write a poem which took the *Comedy* as its model, but out of a sense of poetic justice and family pride, he would show that he too held hell – Dante's hell – in great contempt. And he would do so with the help of Ireland and the Irish.

Fazio's journey in the *Dittamondo* is a journey through *this* world. It begins in Rome, under the guidance of Lady Rome herself. From Rome, with Solinus now in charge, it proceeds through the rest of Italy and then overland to Greece, where the travellers relax on Mount Parnassus, the mountain sacred to Apollo and the Muses, visit the palace of Alexander the Great, with its exquisite statues, climb Mount Olympus, and tour the islands. Thereupon they find themselves in the very north of Europe, marvelling at the Hyperboreans and the Scythians, 'barbarian peoples, bestial, monstrous and wild, just like the monkeys who live in trees', after which they visit Scandinavia and central Europe, travel on through France, England, Scotland, Ireland and the islands beyond, including Thule, and conclude their tour of Europe in Spain.[8] From Spain they cross over to Africa, where they encounter more animals than they do people, and thence to Asia, of which they visit only Babylonia, Egypt and Judaea, for the work, as we said, is incomplete. In a sense, therefore, the journey in the *Dittamondo* resembles the Mediterranean-centred tour of the *oikoumene* in Strabo's *Geography*, but of course Fazio could not have known or read Strabo's work, and in any case the spirit in which the *Dittamondo* is written is that of quite a different age, different even from the triumphalist spirit fostered by Renaissance humanists.

Where Strabo, and the humanists, write in praise of a heroic Rome, Fazio's Rome is a has-been. The Lady Rome who shows him around the eternal city is a fallen lady, in tatters and dishevelled, who plaintively laments the reduced circumstances in which she finds herself and obsessively recounts her decline, spending more than an entire book of the six books of which the poem is composed recounting the history of Rome, that is (since God's creation was ever destined to be Roman), the history of the world, from the days of Noah to the year 1356. And although the hope is that Rome will rise again, the resurgence – Rome's and that of her followers – is more akin to a resurrection than to a renascence. Nothing proves this more convincingly than the thoughts that come to Fazio's mind as he wanders through the wilderness of Africa:

> our soul and hope must fasten fully
> on Him that is supreme goodness,
> fleeing vice and doing good.
> And I, who am in parts so distant
> and midst people so vile and spiteful,
> turn to Him for mercy and forgiveness:
> and, as far as I can, devout and humble,
> I pray Him that He help me on my journey
> and that at last I find myself within His sheepfold.[9]

In the same place, Fra Mauro would be thinking not about the mystery of the world to come, but about the many mysteries which *this* world still held in store for us.

Nevertheless, for all that Fazio's hopes are 'fastened fully' on the otherworld and for all that the primary goal of his journey is, or so he says, the gaining of 'virtue', it is 'knowledge' of the herebelow which constitutes the main focus of the text. The journey, though, is a fictitious one, for as far as we know in real life Fazio never ventured beyond the borders of central and northern Italy. His knowledge too, therefore, is that of an armchair traveller and, with a guide even less all-knowing than Dante's Virgil, Christian though he is made out to be (where Dante had been at pains to deny his own beloved guide the gift and benefits of Revelation), the truth he speaks is often equivocal. This is certainly the case with regard to Ireland.

Silvestri is right. Fazio does give the impression that *Ibernia* is but a part of *Irlanda*, or possibly even that the two are separate places. First he reaches *Irlanda*, but it is *Ibernia* he is really keen to see, so keen in fact that he is prepared to turn a deaf ear to the warnings of his reason: 'we crossed over to Ireland, which among us is worthy of fame, for the noble serge it sends us. Hibernia now awaits us here and beckons us and, even though sailing there is risky, reason was here overcome by yearning'. When he gets there, it is his reason which initially appears to have been right. *Ibernia* – which is one hundred and twenty miles long and derives its name from the 'Iberian ocean' that surrounds it – is a disappointment: 'the people look wild' and the 'countryside harsh'. Very soon, however, his yearning is fully vindicated, as he discovers that there is 'always a gentle and soothing breeze, as in spring', that there are 'clear springs and beautiful rivers', and that 'the earth is so good', the 'great pastures so full of grass' that Ceres, the goddess of harvests and plenty, has 'nothing to reveal of her art'. The land, in other words, does not need to be worked, as food, especially meat and milk, is plentiful. Better still, the inhabitants of this earthly paradise are quite as Petrarch had predicted: they are 'sweet' [*dolce*]. Ireland – Hibernia – was indeed a go-there-what-are-you-waiting-for kind of place.[10]

Whether Fazio and Petrarch were acquainted with one another and had exchanged opinions on Hibernia, we do not know. Neither can we tell for sure who of the two first came to the conclusion that the Irish were not the barbarians tradition had made them out to be, but were 'sweet' and 'happy'. What we can say, though, is that they both reached that conclusion starting from the same premise, namely Gerald of Wales – quite possibly, in the case of Fazio, by an indirect route, that is, via the *Polychronicon* of the English monk Ranulf Higden (d. 1363), an encyclopaedic description of the world and chronicle of world events, from the time of Creation to the early fourteenth century, which contains a chapter on Ireland derived almost entirely from Gerald's *Topography* and which was to rival the *Topography* in popularity throughout Europe, with no fewer than one hundred and eighty-five pre-1500 manuscripts of the work surviving.[11] What is for sure is that Fazio's Ireland is essentially Geraldine in nature, noteworthy above all for its marvels. It has a lake – that 'so pleases me, that my eager eyes still feel delight' – in which wood is changed to iron and to stone yet remains unchanged above the water; another lake that changes cornel-wood into ash and vice versa; a small island, 'where man can never die, but when he is about to pass away, throws himself outside'; remote caverns where no bodies ever decompose, 'so temperate is the air that strikes them'; and pearls, jet, and iris that in the sun produces a rainbow.[12]

However, there is one feature of the *Dittamondo*'s Ireland which is almost entirely absent from the *Topography*, and it is that above all which most tantalized Fazio's 'yearning'. Having marvelled at the island's miracles, he and Solinus then make a special point of enquiring about 'a very holy and devout monastery' and, having been shown the way, 'there we took ourselves and there was where we lodged', until 'those good friars led us to the well that brings fame to the blessed Patrick'.[13] Such was the fame of the place, it would seem, that the author felt no need to explain to his readers what the well was about, but instead simply let them wonder for themselves, by means of an intriguing little dialogue, full of Dantean echoes, whether it was right, in a place like that, not to resist the suggestions of one's yearning. As he and Solinus stand upon the threshold of the well, Solinus – who, like Dante's Virgil when he too is out of his depth, has lost his nerve but is minded not to take the blame upon himself – turns to Fazio and asks: 'What shall we do? Do you wish to cross over into there, you who seek clarity upon every novelty?' For his part Solinus would be happy enough, it would seem, if only 'you' went in, while 'we' stayed here and waited; but to conceal his discomfiture he resorts to irony. Is it really all that wise, he appears to suggest, to look for 'clarity' in a place that is as dark as they come? Fazio, in truth, is none too eager himself. He is excited all right (his excitement communicated to the reader through the use of an anaphora: 'And he said … And I said … And he said …'), but in the end fear has the better of him. Taking a leaf out of his master's book, he too therefore decides to pass the buck, and so replies: 'Without the advice of these friars, I will not go in, for I am made too gloomy by the thought of searching hell right to its very centre'.[14]

The response which he gets from the friars does not do much to boost his confidence. 'Unless you are clean and pure, constant and full of faith', one of them informs him, your safe return cannot be guaranteed. Indeed, the odds of a safe return, the friar further lets him know, are actually rather limited:

> Of Patrick and of Nicholas it is known, without any doubt, that they descended and returned through this gulley. As to the others, I can tell you that of one hundred that are famed for having done so, not one has passed through here; I for sure do not know of any.

For Fazio that is quite enough. He will go no further. Yet to conceal his own discomfiture, Dante-like (as Virgil and Dante proceed through hell and purgatory, their roles are gradually inverted), he now turns the tables on Solinus, announcing solemnly, and with the same degree of irritation parents use when they blame each other for the misdemeanours of their children: 'Solinus, leave this thought, and do not try to tempt *your* Lord' – it is just as well Solinus has been Christianized! – 'I'll be damned if anyone dies on me at this point; let us be content with searching that which is above'. 'Well said', replies Solinus, hardly able to conceal his relief. And in haste the reluctant pilgrims bid farewell to the friars, and 'leave them be'.[15]

It is not so much the friars, of course, whom Fazio wishes to 'leave be'. It is Dante. By turning his back on 'the well that brings fame to the blessed Patrick' (i.e. St Patrick's Purgatory, about which we will find out more in the next chapter), Fazio was in fact turning his back on the possibility of a 'mad flight' like Dante's; and by hinting that there was little 'clarity' to be had from a 'gloomy' place, he was querying the veracity of Dante's account and suggesting that it was better, all told, to hold Dante's hell in 'great contempt'. If hell was so 'gloomy', Dante cannot possibly have seen what he claimed to have seen, and so the 'magnanimous' Farinata might not be damned after all. More importantly, writers who lived in the shadow of Dante were not condemned for evermore to labour under the burden of his genius. 'I am not Aeneas', Dante had cried out when he had been called upon by his master Virgil to enter into hell (evoking the passage in Virgil's *Aeneid* where Aeneas goes to the underworld to find his father).[16] 'I am not Dante', Fazio seems to answer back, 'nor do I want to be. Mine is to search what is *above*'! Fazio's trip to Ireland, in other words, is the occasion for him to proclaim his originality, loudly and clearly, and at the same time to cast aside his authorial *Angst*. No wonder his Irish are so 'sweet'.

To call the Irish 'sweet' was no mean compliment. To modern ears sweetness rhymes with tweeness, but not to ears brought up on Dante. In order to distinguish his own and his friends' style of poetry from what he held to be the unrefined style of his predecessors, Dante had labelled the new poetry 'sweet' and done so with such effectiveness that to this very day it is known as the 'sweet new style' [*dolce stil novo*].[17] Henceforth sweetness was to be the hallmark of the refined and the civilized. However there was better still to come for the Irish – or some of them at least. Thanks to Dante they were to be given the ultimate accolade of being dubbed Florentine.

* * *

There was no prouder Florentine than Dante. His *Comedy* is a passionate *cri d'amour* for Florence. But Florence had not responded in kind. In 1301 Dante was exiled from his native city, for political reasons, and for the rest of his life he was forced to wander from court to court, in northern Italy, throwing himself upon the mercy of generous, or not so generous patrons. It was while in exile that he wrote his masterpiece, and it was in exile, in Ravenna, that he died and was buried. Not long after his death, however, as his fame began to grow, not least owing to the efforts of his sons, who set in train what was in due course to become a veritable cult of their father, Florence tried to get him back, if not in actual fact at least symbolically, and once again to make him Florentine. At about the same time as Fazio was composing the *Dittamondo*, the Florentine authorities charged Boccaccio with delivering annual public lectures on the *Comedy* and in due course, as civic humanism became the creed of the city's ruling class, Dante was given praise for what was now recognized to have been his patriotism and thus officially promoted as a model for other citizens to imitate. To such an extent was he turned into a public figure that he inevitably fell prey to the machinations of Florence's party politics, with opposing factions attempting to lay exclusive claim to his name and legacy, sometimes violently. In 1431 Francesco Filelfo (1398–1481), a leading intellectual who was allied to one of the factions, was knifed and permanently scarred after delivering a public lecture on the *Comedy* and forced to leave the city in fear of his life.[18]

The high point in the officially sponsored cult of Dante occurred at the time of Lorenzo the Magnificent, the then head of the Medici family, who, thanks to the power they derived from being Europe's wealthiest bankers, were in the process of transforming themselves from private citizens into princes, and Florence from a republic into their own private fief. They had many supporters, not least among the middle and lower orders of Florentine society, but of course they also had many enemies, and in 1478 a family of rival bankers, the Pazzi, entered into a conspiracy with the pope, Sixtus IV, who had personal reasons to fear the Medici, in order to assassinate Lorenzo and his brother, Giuliano, and thus rid the city of its overbearing bosses. Giuliano was murdered, but Lorenzo escaped and so the conspiracy failed. Ultimately it only served to deliver even more power into the hands of the Medici, as they ruthlessly set about destroying those who had dared to oppose them, but in the short term it left them vulnerable and isolated, with the pope excommunicating Lorenzo and his acolytes, placing an interdict on the city and even going to war against it. To turn the situation around, Lorenzo therefore had to fight on many fronts.[19]

One of these was the front opened up by the newly invented art of printing. Lorenzo waged a well-orchestrated propaganda campaign, sponsoring the publication of numerous pamphlets in defence of himself and his actions. But nothing rivalled the impact of the publication, in 1481, of an edition of Dante's *Comedy* with an introduction and commentary by Lorenzo's protégé and former tutor, Cristoforo Landino (1424–92), patriotically entitled *The Commentary by the Florentine Cristoforo Landino upon the Comedy of the Florentine Poet Dante Alighieri* [*Comento di Christophoro Landino fiorentino sopra la Comedia di Danthe Alighieri poeta fiorentino*]. Mixing politics and scholarship in a manner not untypical of the time, Landino had been an active member of the Guelph party (Florence's only recognized political party) and was a professor at the university of Florence, where he had lectured for a number of years on classical authors as well as on Dante. His commentary is the fruit of those lectures and had been in the making for a good while. In accordance with the then current vogue for the philosophy of Plato, promoted by Lorenzo himself, it gave by and large a Neoplatonic reading of the work, according to which poetry was the product of divine fury (or in other words, of a trance) and the poet a divinely inspired prophet. Dante thus became the 'divine Dante' and the *Comedy* a message from God.[20]

However, the circumstances demanded more – and less. They demanded that Dante's message be marshalled once and for all to the promotion of Florence's temporal ambitions, that it be brought down to earth, as we said. Florence was now in great danger and it was no longer enough for Dante to be divine. He also had to be unmistakably Florentine: the city's saviour to inspire others to rally to its cause. And all the more so as there were those who were actively attempting to deny him his rights of citizenship, and Florence the right to the status to which it aspired. In 1478 a Milanese publisher had brought out an edition of the *Comedy* which not only sought to commandeer the work for the service of princes in northern Italy but also enveloped it in a weighty commentary (the updated commentary of the fourteenth-century Bolognese Iacopo della Lana) which spoke unflattering words about the Florentines. 'They are not content', says Lana

> with common riches and the normal size of cities but appear to want to mimic Troy. They are not content with normal buildings but want them to be of decorated stone. They are not content to be dressed like men but mimic women's clothing, using quantities of cloth and wrapping themselves up from head to toe. They are not content with normal food but allow themselves all the excesses and extravagancies that money can buy. They are not content with their own lands but have wanted to lord it over all of Tuscany.

Worse still, Lana's commentary was commended to those who would read it for being written in what the publisher claimed was the universal and the most beautiful tongue of Italy, namely Bolognese. 'Iacopo della Lana's commentary', the publisher writes in the introduction,

> was deemed to be the best of all the commentaries available, because it is written in our mother tongue and the tongue of Bologna. For that great city is in the very navel of Italy and thus, through constant interchange, its speech is made up not only of elegant words but of words which are common to all the regions of Italy; and it is no less pleasing and dignified in Italy than Laconic used to be in Greece.[21]

Clearly, something had to be done.

In the best of circumstances Lorenzo and his friends, who were themselves engaged in an active campaign to promote Florentine as the most perfect form of Italian, would have found such claims intolerable; but now the Pazzi conspiracy and its aftermath made them doubly so. Landino thus composed an introduction to his commentary which put paid to them in no uncertain terms, as well as to the claims of those who would challenge Florence's pre-eminence. He did so too in the speech which he delivered at the time of the launch of the *Commentary*, a launch which must have been then, as it would be now, the envy of any other writer. Paraded publicly through the streets of Florence, the work was brought in procession to the Palazzo Vecchio and solemnly presented by the author to the rulers of Florence (the official rulers, that is, not the Medici, who at that point were still the government's cash-in-hand back-seat drivers), while he invited them, in his speech, to 'congratulate your most splendid fatherland on which Almighty God has bestowed such a gift'. Dante, he went on to say, was now being 'restored to his fatherland after a long exile' and henceforth would 'be recognized as being neither Romagnol nor Lombard, nor of the dialects of those who have commented him, but merely Florentine'. Better still, under Dante's god-like and prophetic wings, Florence's empire, as it was commonly referred to in the political discourse of the day, would grow and prosper and the Florentines would be universally acknowledged as constituting 'the fifth element in human matters'.[22]

The unashamedly patriotic introduction then goes on to prove the point by listing all the achievements of the Florentines, in every field of human endeavour. The Florentines, it argues, are quite simply the best, at everything: the best soldiers, the best scholars, the best religious, the best orators, the best musicians, the best painters, the best sculptors, the best lawyers, the best merchants. And they have showered their gifts not only on themselves but on peoples far and wide. Even on the Irish:

> There were in England – Landino claims – three brothers, Gherardo, Tomaso and Maurizio, of the very ancient family of the Gherardini of Florence, sent into exile because of civil strife; these men, in the acquisition of Ireland [*Hibernia*], an island not very much smaller than England, lent the English king such eager, brave and faithful support that when the island had been subdued, they were invested with the lordship of the entire flat part of it, where there are many peoples; and in our own day their descendants remain there as lords, especially the Earl of Kildare and

> the Earl of Desmond, and they have changed neither the name nor the coat-of-arms of the ancient Gherardini.[23]

In other words, the Norman invasion of Ireland was actually a Florentine invasion, and originally the Fitzgeralds were not Norman soldiers but Florentine merchants (the episode is recounted in the section of the introduction where Florence's commercial achievements are hailed). Theirs, moreover, was a true Florentine success story, emblematic of the Florentines' ability, as demonstrated with such brilliance by the author of the *Comedy*, to overcome adversity and be drawn to ever higher planes. Like Dante the Gherardini-Fitzgeralds had been the victims of civil strife and been banished from their native city, and like Dante they had turned misfortune to their advantage and gained such fame that Florence could now proudly acknowledge them as her own. More to the point, like Dante they had been metamorphosed by their excellence. Dante who, because he had chosen to write in the vernacular, had often been considered the poet of the uneducated (why, *even* women could understand him!), was being raised, through the agency of a renowned classical scholar, such as Landino was, to the level of a prophet. The Gherardini, who had begun life as simple merchants, were, by the king they had served with such great chivalry, elevated to the peerage and made lords of many peoples.

It is true, of course, that the Anglo-Norman settlement of Ireland, to the extent that it was made secure by the financial backing of Tuscan bankers (more Lucchesi, in reality, than Florentines), was an Italian settlement, but where the urban myth of the Gherardini-turned-Fitzgeralds takes its origin is none too clear. The imperializing hubris which lies behind it is present to an extent in Dante, but it is the humanists, and above all the civic humanists, not to say the civic humanists of Florence, who endowed it with respectability (Dante's imperialism, like Fazio's for that matter, is more nostalgia for a lost golden age than triumphalist zealotry, as is the humanists'). One of its more famous promoters was Leonardo Bruni (1370–1444), a leading humanist who in 1427 became Chancellor of Florence, a post previously occupied by his master Coluccio Salutati; and according to a *Book of Memoirs of the Family of the Gherardini of Florence* supposedly written in 1585 by one Canon Nicholas Gherardini and apparently in the possession of the Fitzgerald Dukes of Leinster as late as the nineteenth century, it is Bruni who provides one of the earliest pieces of documentary evidence regarding the Gherardini-Fitzgerald connection.[24]

According to Canon Nicholas, in the 1430s, some twenty-five years after an Irish bishop had passed through Florence and together with him 'an Irish priest, of the cathedral church of Ardfert, called "Maurice", who was of the family of the Gherardini settled in that island' and who 'at once acknowledged us for his relations', Bruni wrote a letter of introduction to the Earl of Desmond on behalf of a member of the Gherardini family travelling to Ireland. It would seem, however, that in doing so he was less interested in renewing family ties than he was in consolidating Florence's imperial destiny. His letter reads as follows:

> If the assertion which circulates about you is true, namely that your ancestors were of Florentine origin [...] we rejoice [...] and we take pride in the fact that citizens of ours held great lordships not only in Apulia, in Greece, and in Hungary, but that through you and your relatives, Florentines also rule over Ireland, which is the furthest of the islands. Oh! how great is the glory of our city, oh! how exceptional God's kindness to our people, that so many noblemen, so many lordships were scattered throughout the whole wide world![25]

It would even seem, in fact, that Bruni himself was none too convinced by the story – '*if* the assertion if true ...' – and we ourselves must inevitably wonder whether such a letter actually ever existed, just as we must set aside any notion (because the dates will not allow otherwise) that the Irish bishop who passed through Florence with Father Maurice from Ardfert might be Silvestri's Bishop Milo. What is for sure is that there is not the slightest reference to Ireland in works of Bruni's where one might quite possibly have expected to find some.

Ireland is mentioned neither in Bruni's *Panegyric of the City of Florence* [*Laudatio florentinae urbis*], a eulogy written in praise of Florence's ambitions to be free and bring its freedom to others (i.e. to visit its imperial destiny upon its neighbours), which is said to have been a model for Landino's introduction to the *Commentary*, nor in his *Funerary Oration for the Florentine Knight Nanni Strozzi* [*Oratio in funere Nannis Stroze equitis florentini*], in which he eulogizes the destiny of Florence even more explicitly and where, if the fortunes of the Gherardini really were the cause of such rejoicing, to him and his fellow citizens, it is hard to imagine he would not have wanted to recall them, given what he does recall:

> If it is Britain you seek out, well nigh the furthest island in the world, situated in the ocean, you will come across a great multitude of our fellow citizens; and if it is France or Germany you wander through, there too, where least you expected it, you will find very many of our citizens. And what should I say of Paris, what of Barcelona, what of Avignon?[26]

And what of Ireland? – we are bound to ask, as we bear in mind the words of Canon Nicholas. Yet it is as though Ireland did not even exist for Bruni, as though it were 'out of this world'. And from this we can only deduce that the legend of the Florentine Fitzgeralds probably originated much closer in time to Landino's *Commentary*, possibly in the *Commentary* itself.

That, though, is not really what matters. What is more important is that, thanks to Landino and the on-going appeal of Dante's *Comedy*, there were few among Italy's reading élite who would not have known about the legend. Despite its chauvinistic conception and the domestic concerns which marked its birth, Landino's *Commentary* was a run-away success. The twelve hundred copies of the first edition (a truly remarkable number for an academic book, by any standards) having soon sold out, the work was reprinted no fewer than fifteen times over the course of the following one hundred years (none of them, ironically, in Florence) and for over half a century remained the standard edition, throughout Italy, of Dante's *Comedy*, or, as it was henceforth to be called, thanks to the effectiveness of Landino's deification of its author, of the *Divine Comedy*. Everybody who was anybody – and by all accounts the *Commentary* was read by many luminaries – would thus have heard about the Irish saga of the Gherardini of Florence.

* * *

There was somebody, however, who would have known about it even if Landino's *Commentary* had not been the success it was. He was a former pupil of Landino's, by the name of Francesco Berlinghieri (1440–1501), who also belonged to the Medici circle and shared with them all the interests of Florence's wealthy middle class. From a family of merchants, Berlinghieri received a humanist education, as was now the rule in that class, dabbled in politics, as was the class's prerogative, wielded the pen, as many of his peers also did, and all the while cultivated the good will of the city's leading family, often working for them and their minions, and often playing with them, in particular in the Platonic Academy, a gentlemen's club, so to speak, which promoted the cult of Plato and sought to give Florentines all the wisdom and refinements (and *imperium*-worthiness) of ancient Athens. Berlinghieri, as would have been expected, was also an ardent admirer of Dante. His admiration, however, took him to lengths no-one else would have dared to go to. Wishing to feed his contemporaries' interest in geography, besides gifting a superior manuscript of Ptolemy's *Geography* to Lorenzo the Magnificent, he spent close on twenty years of his life transforming the work into a 'mad flight' like the *Comedy*, in due course bringing out, one year after the publication of Landino's *Commentary*, and through the offices of the same printer, a massive tome entitled *The Geography of the Florentine Francesco Berlinghieri Produced in Terza Rima and the Tuscan Language [...] according to the Geography and the Maps Produced by Ptolemy* [*Geographia di Francesco Berlinghieri fiorentino in terza rima et lingua toscana distincta [...] secondo la Geographia et distinctione de le tavole di Ptolomeo*].

The flight was mad indeed. Transposing Ptolemy into *terza rima* is about as feasible, and as inspirational, as translating a telephone directory, including the introductory section on how to avail of operator services, into iambic pentameters. Unsurprisingly the reading public was not impressed. While Landino's homage to Dante sped to the top of the best-seller list, Berlinghieri's never made it to the bottom. As far as we can tell, hardly anyone bought the book, and forty years later unsold copies of it were still being remaindered by his heirs. Berlinghieri's idea, though, had certainly been an original one. When he started working on it, in the 1460s, no edition of Ptolemy's *Geography*, in either Greek, Latin or any modern translation, had yet been published. It was to be another ten years before the Latin version first saw the light of day (in Vicenza, in 1475), a further two before a (Latin) edition of the work was published together with any maps (Bologna, 1477), and the edition of the work which is regarded as the best executed of the early printed ones, better still than the splendid one dedicated to Lorenzo's nemesis, Pope Sixtus IV (Rome, 1478), came out in the very same year as Berlinghieri's poem (Ulm, 1482). The Greek original was not to be published until 1533, and a proper Italian (prose) translation only came out, in Venice, in 1548. In addition Berlinghieri's is the first edition to add 'new' maps to the twenty-seven so-called 'old' Ptolemaic maps, and it contains the earliest printed modern maps of Italy, Spain and France. Because of this, Berlinghieri has been hailed as an innovative geographer, though it would in fact appear that his text and the maps actually found their way into print by different routes, and that Berlinghieri may not have been responsible himself for the inclusion of the maps in his *Geographia*. In many instances there is a mismatch between the maps and the text, and any reader trying to locate on a map a place that is named in the text will often find it more impossible to do than actually making sense of the text, which is itself a formidable task, given the idiosyncrasies of Berlinghieri's Tuscan and the many printing errors which the work contains.[27]

Equally as problematic would be any attempt to make the *Geographia* of Berlinghieri tally perfectly with the *Geography* of Ptolemy, for Berlinghieri's work is of course a free adaptation of Ptolemy's, as it could only be, since it is discursive and in verse where Ptolemy's is tabulated and in short hand, and since Berlinghieri often adds in extra information taken from different sources. What is more, Berlinghieri's work tries to tell a story, by turning the raw data found in Ptolemy, including the theoretical discussions from the early chapters of the work, into the stuff of good-for allegory. The work is thus constructed as a journey undertaken by the poet-pilgrim (Berlinghieri) under the guidance of a sage from antiquity (Ptolemy, of course), and it reads like a kind of *Geographic Comedy*, in which the guide, like Dante's Virgil, or Fazio's Solinus, does not simply show the way but seeks to lead the pilgrim unto higher plains of enlightenment, although the enlightenment, it must be said, often leaves a lot to be desired. The guide, indeed, frequently sounds like one of those guides you find yourself having to double-guess on exotic holidays: he is well meaning all right, but you cannot help but wonder whether, despite his best assurances, it really is the same language as you which he says he speaks. More often than not, the language spoken by Berlinghieri's Ptolemy is bewildering beyond belief. In English it can sound (and look) a bit like this: 'And now admire *Regia* &

here is *Beba* & *Labero*, thother *Rhegia* & *Macolico Arferdense*, & *Duno*'. All of which, naturally, are in Ireland.[28]

Still, despite the guide's shortcomings, the work is all-inclusive. It takes in all the world, as it was known to Ptolemy and Berlinghieri, and it shies away neither from islands that are deemed to be un-Christian nor from islands that might be said to be not fully European. To tour this world it takes a full eight days, and the tour begins with an invocation to Apollo, which causes Ptolemy suddenly to appear from behind a cloud in order to announce to the would-be pilgrim, in true Dantesque (or possibly, Fazioesque) fashion, 'I will be your leader and your guide and the indicator of the universe which you are searching, returning you at last to your dwelling'. The pilgrim, though, must first stay put and listen while his guide, 'in whose wake no one ever errs', explains the difference between cosmography and chorography, as well as how to measure the earth, how to calculate meridians and parallels, how to draw maps, and other matters of the sort. Only then do the travellers set off. Their wanderings begin in Ireland, for 'Claudius Ptolemy, on the second day, sets the islands of Hibernia and of Albion, and tripartite Spain at the end of the world'. From there they move on to western Europe, then to northern and to central Europe, and then to Italy, where they remain for quite a while, before continuing on their way via eastern Europe to Greece and Crete, where the first part of their journey ends. From Crete they sail to Africa, and from Africa in due course they proceed to Asia, which they cross all the way from Asia Minor to 'India beyond the Ganges', bringing their tour of the world to a close in Taprobane.[29]

The tour of Ireland, which takes the wayfarers from the north of the island to the west, to the east, to the south, to the west again, to the south again, and then to the centre, following the demands, presumably, of *terza rima*, begins as confusingly as it continues. 'This is now the British island of Hibernia [*Ibernia*], that first of all appears to us', Ptolemy explains, 'and by the inhabitants Ireland [*Irlanda*] it is called. Named Hibernia after winter, and British because the Bretons owned it, to whom Edward now commands, Armorique [i.e. Brittany] by settlers it is also termed'.[30] On top of these eccentric appellations and this unusual mix of settlers, the island also has a whole array of different tribes, with exotic-sounding names – the

Robogdii, the *Vennicni*, the *Darini*, the *Usluntii*, the *Ebdani*, the *Cauci*, the *Manapi*, the *Coriondi*, the *Briganti*, the *Vellabri*, the *Erdini*, the *Autini*, the *Gangani*, the *Herpeditani* – and intriguingly it is also home to the 'errant Irish'. In addition it features a good few rivers, capes and cities, as well as '*Duno* beneath the Gherardine empire', a 'holy cave', and 'many islands' inside a 'fortunate lake' which are said to be 'worthy not to be forgotten', above all because of 'the great virtue that rains down on one of them, causing its inhabitants never to be led to death until they are removed from it'. The names of the rivers and the capes, like those of the tribes, are mostly drawn from Ptolemy. Many of the cities, though, derive their own from a more recent source – a portolan, it has been suggested: a map, that is, of ports and coasts, such as sailors and merchants liked to use – and to modern ears not all of them sound quite so unequivocally alien. There is 'the city of *Dregoda* once termed *Eblana*'. There is 'the city you now call *Donueli*'. There is *Corcha*, *Rois*, *Alameriche*, *Arferdense*, *Caselle*, *Armachan* and the city *Dublinense*.[31]

If the readers of the time shied away from accompanying Berlinghieri on his mad flight and we ourselves can only praise them for having done so, given the doubtful 'virtue' of his *Geographia* and the garbled 'knowledge' which it proffers, Berlinghieri himself appears not to have been aware of its shortcomings. True, in letters of dedication of the work, he does seek forgiveness for 'the harsh strangeness and the unusualness of the barbaric names' it contains and for any 'errors in the text or on the maps' which the printer and engraver may have omitted to correct, but in the introduction he displays not the slightest hesitation regarding what he sees as its benefits, and the benefits to be derived more generally from the study of geography. 'Whoever is enlightened by concern for himself, whether he rules in private or in public, fares ill if he does not learn this science', he pronounces, 'for just as the sun is more resplendant than any other light, so whoever acquires this knowledge shines brighter than any other man'.[32] Presumably this conviction is the expression not only of Berlinghieri's own belief, but of the feeling quite common among humanists that, for all that Ptolemy's was the geography which held the key to the future of the science, it was Strabo's which contained the greater truth. Geography, as we said, was not just about measuring the world, it was about taking the world's measure. It was about what mattered to people who mattered. It therefore had to be *moralisé*, Strabonified. That is what Berlinghieri's *Geographia* is. It is Ptolemy Strabonified, as well of course as Dantified, to stress its sense of purpose, and in the process to present an Ireland which, for all its 'marvels', was fit for 'governmental purposes' – while also satisfying the reader's 'concern for himself', to use Berlinghieri's own words, or 'care for the self', to use the previously quoted modern equivalent.

Although it is hard to credit, that sense of purpose, insofar as it applied to Berlinghieri's work, was also shared by Florence's two foremost arbiters of taste, Lorenzo the Magnificent himself and the high priest of Neoplatonism, Marsilio Ficino (1433–99). In order to ensure that those who mattered would pay attention to his work, Berlinghieri had intended to place it under the protection of the most powerful ruler of the day, namely the Turkish sultan Mohammed II (1432–51–81), the conqueror of Constantinople, for whom, 'because of his many outstanding and almost incredible virtues', 'because of his very keen interest in the subject', and 'because he ruled over the largest part of the universe', it seemed to be intended as of right. Unfortunately, Mohammed's 'most happy soul' had departed from 'this life of woes' before the work had been brought to print, and since Berlinghieri did not know whether a work of the sort would appeal to his son, he had decided to offer it instead to 'him whose knowledge and whose mastery of arms nowadays is considered by all peoples to be equal to that of ancient military leaders', that is, to Federigo da Montefeltro, Duke of Urbino. However, for Berlinghieri's 'bad luck', or rather, 'by the will of God and as a marvellous and true sign and a good omen of the greatness of the empire' of Mohammed's heir, Federigo himself had died, only a few weeks after the work had been published, and though Berlinghieri had then toyed with the idea of dedicating it to Federigo's son, Guidobaldo (1472–82–1508), in the end, because he had been assured that the gift would be welcome, he had made up his mind to dedicate it to Mohammed's son after all, that is, to Sultan Bayezid (1447–81–1512).[33] And just to be on the safe side, lest fate thwart his plans once again, he had also dedicated a copy to Bayezid's rival sibling Djem (1459–95), a hostage in Savoy to the machinations of his fearful brother, who paid good money to ensure that Djem would not return to Istanbul, and of the western powers who, even as they took the money, conspired to make Djem a Christian and place him on the throne of Istanbul instead of Bayezid.

Just why Berlinghieri's work was made a pawn in the power politics of the day is a story too long to be recounted here. Suffice it to say that the person who advised him on the matter was the long-time agent of the Medici in Constantinople, Paolo da Colle, and that da Colle was then appointed to be the bearer of the gift, delivering the *Geographia* personally to both Bayezid in Constantinople and to Djem in Savoy. What prompted Lorenzo, who was no mean poet in his own right, to wager his reputation as an artist on a poem so totally devoid of artistic merit, and whether or not the recipients of the gift were in a position to estimate its value, must remain a mystery. What is for sure is that both Berlinghieri and Lorenzo failed to ensure that the mouth of the gift horse, so to speak, was given a good wash. Not only are the dedications to Bayezid and Djem simply stuck onto the frontispiece of the respective presentation copies with a bit of glue, but, as far as we know, both copies still contain the original dedication of the work to Federigo da Montefeltro composed by Marsilio Ficino.

Ficino had only just given his blessing to the *Commentary* of Landino, by means of a 'letter of congratulations', written in both Italian and Latin, which is included in the introduction and in which he rejoices at the splendid manner in which Dante, thanks to Landino's endeavours, had been brought to life again and returned from exile to his native city. In the case of Berlinghieri, though, Ficino goes even further, stating grandiloquently:

> To him whom almighty Jupiter, master of the universe, created to have power [*imperium*] over the earth, to him whom Pallas and Mercury, children of Jove, long ago made worthy of such power, to him, rightly, do both the academy that worships these deities dedicate this Platonic model of the earthly kingdom and our academic friend, Berlinghieri, gift this full poetic representation of the world, at a time when, with biddings of good fortune, states, leaders and kings again and again do delegate to him both the spear of Pallas and the club of Hercules, so that they be the mighty mistress of Italian wars, even as, almost to a man, the powerful decree that Italy at war should obey Duke Frederick, the ever unbeaten. To him meanwhile this academy and this academician give up the entire world in both peace and war.[34]

Power throughout the universe, empire to the ends of the earth, victory in war and peace, right and might, the favour of the gods, the ambitions of dukes and kings, the glory of Italy: all the hubris Platonists, and humanists, could muster is here, in these few short sentences. And the message could not be simpler: from Florence and her Platonic Academy there came the means (in this instance geography) of subjecting the whole world and making it perfect. For sure, such flights of fancy were a heavy burden to place upon the shaky shoulders of Berlinghieri's limping verses but, together with those very verses, they do give us a precious and memorable insight into the way 'representations of the world' were evolving.

Representations of the world – or in other words, geography – helped to set the limits of this world, establishing what was in and what was out of it. They helped to distinguish the 'barbarians' from the people who were 'sweet' and 'happy', establishing what was 'amazing' and what was not, who was Other and who was not. They helped to make sense of the otherworld, sanctioning when it was best to 'go there' and when it was preferable to 'leave it be', or harnass it to the service of the herebelow. But above all they helped to promote *imperium*, and establish who the world belonged to and who could have no claim upon it. As such they presupposed that any empire, even an Infidel one, was better than no empire, and that every land and island, even out-of-this-world Hibernia, had to be part of some empire, and thus, if needs be, could be conquered, whether by the Romans, the Turks, the Florentines, or the Fitzgeralds.

However, not yet everybody, it must be said, was all that keen on empire-building. Even as Landino and Berlinghieri were publicly singing of 'the Gherardine empire', in Ferrara, in northern Italy, Matteo Maria Boiardo (1441–94), the court poet of the city's rulers, was readying for publication his chivalric romance *Roland in Love* [*Orlando Innamorato*], which was to enjoy huge popularity and in due course lead to the creation of an even more popular sequel, *Roland Gone Mad* [*Orlando Furioso*] by Ludovico Ariosto (discussed in the next chapter). The poems recount the martial and amorous adventures of knights and damsels at the time when Emperor Charlemagne, early in the ninth century, was waging war against the Saracens who were bent on defeating Christendom. Not infrequently Charlemagne is in danger of his life and has to be rescued by his valiant paladins. On one such occasion, in Boiardo, his saviour is the English prince Astolfo, and to reward him, Charlemagne makes him an offer which he feels he cannot refuse. But

Astolfo is not impressed. 'Trying hard to retain him, Charles wished to give him all of Ireland. But he is destined to take leave, for Ranaldo and Orlando he must retrieve'.[35] Ireland? – Astolfo seems to think – what kind of gift is *that*?

Notes

1. Rubiés (2000), p. 44.
2. The most famous exponent of the view that the Renaissance was a godless era is Jakob Burckhardt (1818–97), who claimed for instance that 'The nonbelief prevalent in Italy at that time is notorious, and whoever takes the trouble to look for proofs, will find them by the hundred' (Burckhardt (1964), Part 6: 'Morality', p. 321).
3. 'El paradiso de le delicie non solamente ha sentimento spiritual ma etiam quello esser uno luogo ne la terra situado mette sancto Augustino sopra el Genesis et ancora nel libro De Civitate Dei, el qual luogo è molto remoto da la habitation e cognition humana, posto ne le parte oriental segondo la doctrina del sacro doctor Beda' (Mauro (2006), p. 729).
4. 'Non vogliate negar l'esperïenza,/di retro al sol, del mondo sanza gente./Considerate la vostra semenza:/fatti non foste a viver come bruti,/ma per seguir virtute e canoscenza' – 'folle volo' – 'lato mancino' – 'l'altro polo' (*Inf.* 26.118–27: Dante (2001), vol. 1, pp. 389–90).
5. 'Etiopia di sopra, in oriente,/con le selve d'Egitto s'accompagna/e, di verso aquilone, il Nilo sente./Dal mezzogiorno l'Ocean la bagna/e 'n vèr zeffiro tanto si distende,/che porge ad Atalante le calcagna' – 'la qual fra noi è degna di fama/per le nobili sarge che ci manda' (Uberti (1952), vol. 1, p. 27: I, ix, 22–7; p. 328: IV, xxvi, 28). On Charlemont see Anon. (1884), p. 83, Kelly (2006).
6. On Fazio see Cardini (1987), De Anna (1994), Milanesi (1994), Ragni (1986).
7. 'Magnanimo' – 'com' avesse l'inferno a gran dispitto' (*Inf.* 10.73, 36: Dante (2001), vol. 1, pp. 157, 154).
8. 'Popoli barberi,/bestiali, mostruosi e salvatichi/quanto le scimmie che stanno tra gli alberi' (Uberti (1952), vol. 1, p. 281: IV, x, 1–3).
9. 'Dessi fermare l'anima e la spene/del tutto in Colui, ch'è sommo bono,/fuggendo i vizi e operando il bene./Ed io, che 'n sì lontana parte sono/e tra gente sì dispettosa e vile,/ricovero a Lui per grazia e perdono:/e, quanto posso, divoto e umile/Lo prego che m'aiuti nel cammino/e ch'io mi truovi, al fin, del suo ovile' (Uberti (1952), vol. 1, p. 415: V, xxviii, 16–24).

10. 'Passammo in Irlanda/la qual fra noi è degna di fama/per le nobili sarge che ci manda./Ibernia ora qui ci aspetta e chiama/e, benché 'l navicar lá sia con rischio,/la ragion fu qui vinta da la brama' – 'Ibero oceano' – 'Questa gente, benché mostri selvaggia/e, per li monti, la contrada acerba,/non di meno ella è dolce a chi l'assaggia./Quivi son gran pasture e piene d'erba/e la terra sí buona, che Cerera/niente a l'arte sua mostrar si serba./Quivi par sempre, come in primavera,/un'aire temperata che gli appaghi,/con chiare fonti e con belle rivera' (UBERTI (1952), vol. 1, pp. 328–31: IV, xxvi, 28–33, 105, 37–45).
11. See HIGDEN (1865). On Higden see Taylor (1966).
12. 'Che sí mi contenta,/ch'ancor diletto n'han gli occhi miei vaghi' – 'dove l'uomo mai morir non puote,/ma, quando in transir sta, fuor se ne getta' – 'sí temperata l'aire vi percote' (UBERTI (1952), vol. 1, p. 329: IV, xxvi, 47–8, 59–60, 63).
13. 'Un monister molto santo e divoto./Lá ci traemmo e lá fu il nostro ospizio./Poi que' buon frati al pozzo ci menaro,/lo qual dá fama al beato Patrizio' (UBERTI (1952), vol. 1, p. 330: IV, xxvi, 69–72).
14. '"Che farem noi? Vuo' tu passar qua dentro,/che d'ogni novitá cerchi esser chiaro?"/"Senza il consiglio, rispuos'io, non ci entro/di questi frati, ché troppo m'è scuro/pensar cercar lo 'nferno in fino al centro"' (UBERTI (1952), vol. 1, p. 330: IV, xxvi, 74–8).
15. '"Se netto e puro,/costante e pien di fede non ti senti,/se v'entri, del tornar non t'assicuro"' – '"Di Patrizio e di Nicola/è manifesto, senza dubbio alcuno,/che scesono e tornâr per questa gola./De gli altri ti so dir che di cento uno,/che porti di ciò fama, qui non passa:/e io per certo non ne so niuno". "Solin, diss'io, questo pensier lassa/e non volere il tuo Signor tentare;/tristo sarò s'alcun qui mi trapassa;/basti a noi quel di sopra cercare". [...] li lasciammo stare' (UBERTI (1952), vol. 1, p. 330: IV, xxvi, 85–96: emphasis added).
16. 'Io non Enëa [...] sono' (*Inf.* 2.32: DANTE (2001), vol. 1, p. 23). Aeneas's visit to the underworld occurs in Book Six of the *Aeneid*.
17. See *Purgatorio* 24.57 (DANTE (2001), vol. 2, p. 412).
18. For an introduction on Dante in the Renaissance see Haywood (2003a), Salvadori (2003).
19. On the Pazzi conspiracy see Martines (2003).
20. On Landino see introduction in DANTE/LANDINO (2001) and LANDINO (2008); also Gentile (2000) 6.22, Haywood (2004).
21. 'Chissi non si contentano di communali richezze & grandeza di cittade. ma pare chigli voglia contrafare troia. non si contentano di comuni palagi. ma voglionli di pietra chontonata [*recte?* contigiata]. non si contentano dandare in habito virile ma contrafano le feminile in grandeza di panni & in andare stolati. non si contentano di comunale vivanda. ma fano quelle dispiratezze & dilasciamenti

che fare si possono inffolgoratezza. non sono contenti di loro contado. ma hano voluto signoreggiare tutta toschana' – 'Iacobus lanaeus materna eadem & bononiensi lingua superare est visus cum sit illa urbs ita in umbilico italie posita ut assiduo commertio non tersa solum vocabula: sed provintiis omnnibus etiam communia habeat. nec minore gratia dignitateque sit it italia bononiensis sermo quam laconicus olim in grecia fuit' (DANTE/LANA (1478), f. 42r, dedication).

22. 'congratulerete alla vostra splendidissima patria alla quale el sommo Dio tanto dono habbi conceduto' – 'dopo lungo exilio restituito nella sua patria et riconosciuto né Romagnuolo essere né Lombardo, né degli idiomi di quegli che l'hanno comentato, ma mero fiorentino' – 'e Fiorentini essere nelle chose humane el quinto elemento' (DANTE/LANDINO (2001), vol. 1, pp. 111–12, 221, 239). The foremost promoter of Florentine imperialism was Lorenzo de' Medici (Lorenzo the Magnificent). In his (unfinished) commentary to his own poems, for instance, he wrote: 'And perhaps things will still be written in this language which are ingenious and important and worthy of being read; especially as it can be said that hitherto it has been the adolescence of this language, because all the while it is growing more elegant and noble. And in its youth and adulthood it could easily reach even greater perfection; and all the more so if there is added to it some great success and an increase of the Florentine empire, as is not only to be hoped for, but striven for by good citizens with all their mind and might' ['E forse saranno ancora scritte in questa lingua cose sottili e importanti e degne d'essere lette, massime perché insino a ora si può dire essere l'adolescenzia di questa lingua, perché ognora più si fa elegante e gentile; e potrebbe facilmente, nella gioventù e adulta età sua, venire ancora in maggiore perfezione, e tanto più aggiugnendosi qualche prospero successo ed augumento al fiorentino imperio: come si debbe non solamente sperare, ma con tutto l'ingegno e forze per li buoni cittadini aiutare']: MEDICI (1992), p. 120. For more on this see Brown (2000).

23. 'Erono in Inghilterra tre frategli, Gherardo Thomaso et Mauritio dell'antichissima famiglia de' Gherardini di Fiorentia mandati in exilio per dissensioni civili. Questi nell'acquisto d'Hibernia, isola non molto minore d'Inghilterra, sì prompta et sì francha et fedele opera prestorono al re d'Inghilterra, che sobgiogata l'isola furono insigniti della signoria di tutta la parte piana di quella dove sono molti popoli; et anchora ne' tempi nostri restonvi e discendenti signori, et *maxime* el conte di Chaldaria, et el conte di Dicimone. Né nome hanno mutato né insegne de gl'antichi Gherardini' (DANTE/LANDINO (2001), vol. 1, p. 245).

24. On Italian bankers in Ireland see above Chapter Two, note 37. As an introduction to Bruni see BRUNI (1987). On the *Book of Memoirs* and the Florentine Fitzgeralds see Fitzgibbon (1876–8), Gamurrini (1671), pp. 111–17, Kildare (1872), Meehan (1878).

25. 'Si vera est assertio, quae de vobis circumfertur scilicet vestros progenitores fuisse ab origine Florentinos [...] gaudemus [...] ac nobis gratulamur, quod cives nostri non solum in Apulia, et in Grecia, et Ungheria magnas dominationes habuerunt, verum et in Hibernia quae est ultima Insularum per vos et vestros Fiorentini dominantur. O magnam gloriam nostrae civitatis, O singularem benevolentiam Dei erga populum nostrum, ex quo tot proceres, totque dominationes fuerunt per universum orbem terrae diffusi' (Meehan (1878), pp. 246, 243).
26. 'Nam sive tu Britanniam quaeras, extremam pene mundi insulam in oceano sitam, permagna tibi illic aderit nostrorum civium multitudo; sive tu Galliam sive Germaniam lustres, permulti etiam ubi minime opineris cives nostri occurrent. Quid memorem Parisius, quid Barchinonam, quid Avinionem?' (Bruni (1744), p. 4).
27. On Berlinghieri and his work see Skelton's introduction in Berlinghieri (1966); also Almagià (1961), Babinger (1963), Haywood (2009a), Veneziani (1982).
28. 'Et Regia hor mira & Beba eccho & Labero/laltra Rhegia & Macolico Arferdense/& Duno' (Berlinghieri (1966), bk 2, chap. 4).
29. 'Io saro tuo duca & guida & inditio/dell universo il che cercando vai/riducendoti alfine al tuo hospitio' – 'seguendo te che mai chi ti segue erra' – 'Claudio tholomeo il di secondo/Ibernia & Albione isole hispania/tripartita allo extremo pon del mondo' – 'India fora del Gange' (Berlinghieri (1966), bk 1, invocation; bk 2, chap. 1; bk 7, chap. 2).
30. 'Britannica hora ibernia isola e questa/& dagli habitatori e decta irlanda/la quale in prima ad noi si manifesta./Dalla uernata ibernia si domanda/britannica perche era da brettoni/possessa a quali adouardo hor chomanda./Armorica anchor decta da coloni' (Berlinghieri (1966), bk 2, chap. 4).
31. 'Iberni erranti' – '& Duno sotto al gherardino impero' – 'sacro cauo' – 'lago fortunato doue/isole molte degne non doblio' – 'Di loro in una tanta virtu pioue/che gli habitanti a morte mai riduce/per insin che dallei non gli rimuoue' – 'Dregoda gia decta eblana cictate' – 'quella che hora uocate/Donueli' (Berlinghieri (1966), bk 2, chap. 4).
32. 'Alquanto aspra per la nouità et indissuetudine de barbari nomi' (Berlinghieri (1482b)); '... nel texto o nelle tabule fussi alchuno errore dallo impressore facto o dallo intagliatore alla breuita del tempo che non ha lasciato emendare ben che per se corregiere si possi' (Berlinghieri (1482a)); 'Qualunque di se stesso cura accende/o che al priuato o el publico gouerna/mal puo se tal doctrina non apprende: / [...] Che chome el sol piu che ciaschuna luce/splende chosi chi tal doctrina acquista/assai piu che gl altri huomini riluce' (Berlinghieri (1966), bk 1, invocation).

33. 'El tuo severissimo [*recte* serenissimo] patre Meemet innanzi a tutti aparue sì per le sue molte et maxime quasi incredibili uirtù sì per esser di tale facultate studiosissimo sì per imperitare a grandissima parte dello uniuerso allui apparteneua' – 'la sua felicissima alma partì di questa misera uita ne sapendo se a suoi proprii figliuoli tale opera grata si fusse quello che et per sapientia et per disciplina militare a tempi nostri apresso di ciascheduna gente con qualunque di quelli antichi imperatori si poteua connumerare' – 'mia aduersa fortuna' (Berlinghieri (1482b)); 'che iddio habbi uoluto chosi accio che si riseruassi Alla(?) tua(?) Gloriosissima maiesta. chome optima et uera diuinatione et buono augurio della grandeça del tuo i[m]p[er]io' (Berlinghieri (1482a)).
34. 'Quem iuppiter omnipotens orbis totius Dominus ad terreni orbis imperium procreavit: Quem pallas mercuriusque Iovis filii iamdiu tanto fecere imperio dignum huic et academia horum cultrix numinum platonicam terreni regni formam non iniuria dicat: huic et academicus Berlingherius noster universam orbis poeticam figuram eo largitur tempore quo res publice Ducesque et Reges iterum atque iterum Tum Palladis hastam: Tum Herculis clavam concedunt felicibus auspiciis potentem belli italici dominam. Acdum plerique potentes Federico duci semper invicto Italiam bello parere volunt. Interim et academia et academicus eidem tam pace quam bello totum subicit orbem' (Berlinghieri (1966), dedication). Both versions of Ficino's letter of congratulations are to be found in Dante/Landino (2001), vol. 1, pp. 268–70.
35. 'Carlo si forza assai de il ritenire:/Irlanda tutta li volea donare./Ma lui se è destinato di partire,/Ché vol Ranaldo e Orlando ritrovare' (Boiardo (1936), vol. 1, p. 305: I.70).

CHAPTER FOUR

Between Worlds
Imagining Ireland in the Footsteps of St Patrick

Authors and works discussed in this chapter (in chronological order):

- *Treatise on St Patrick's Purgatory* [*Tractatus de Purgatorio Sancti Patricii*] (1184)
- Jacopo da Varagine (c.1228–98), *The Golden Legend* [*Legenda aurea*]
- Checco Rossi da Meleto (fourteenth century), *Letter to the Magnificent Lord Malatesta the Hungarian of the Malatestas, in which He is much Praised, especially for His Journey to the Well of St Patrick* [*Epistola ad magnificum dominnum Malatestam Ungarum de Malatestis, in qua plurimum commendatur, et maxime de itinere ad Puteum Sancti Patricii*]
- Franco Sacchetti (c.1333–1400), *Three Hundred Tales* [*Trecentonovelle*]
- Andrea da Barberino, aka Andrea Mangiabotti (c.1371–1431), *Poor Little Guerrino* [*Guerrino il Meschino*] – also see Chapter One
- Antonio Mannini (fourteenth-fifteenth century), Letter to his brother about his visit to St Patrick's Purgatory
- Ludovico Ariosto (1474–1533), *Orlando Furioso* – also see Chapter Five
- Francesco Chiericati (c.1480–1539), Letter to Isabella d'Este about his visit to St Patrick's Purgatory
- Tullia d'Aragona (c. 1510–56), *The Poor Little One, otherwise Known as Guerrino* [*Il Meschino, altramente detto il Guerrino*]
- *The Acts of the Saints* [*Acta sanctorum*] (1735)

Perhaps Astolfo cared too much about others and not enough about himself to be interested in Ireland. Perhaps he should have been a pilgrim. By definition pilgrims have to be the category of travellers who care most about or for themselves. That is the whole point of going on a pilgrimage: to gain everlasting selfhood. And Ireland offered Christian pilgrims the chance to visit one of the most famous places of pilgrimage in the world, for which Italians in particular showed a certain fascination. However, the fascination could sometimes get out of hand. For some it had more to do with care for themselves precisely – 'because I'm worth it', as a famous television ad has it – than with the penitential benefits of pilgrimage. But that too, maybe, was a way of seeking to establish a meaning outside the sacred.

Be that as it may, within a few years of Astolfo turning up his nose at Ireland, Charlemagne had apparently had a change of mind about the place. Instead of giving the Irish away, he was drawing them to his bosom. The Saracens were now closing in on him and he needed all the help he could get. He therefore dispatched his loyal paladin Rinaldo of Montalbano to muster troops 'from Scotland, and Ireland, and England, and from the islands round about'. The Irish duly responded to the call and, at Rinaldo's command, they gathered outside London, where, together with the other troops, their new commander put them through their paces, before detailing them to the front. The review of the troops was a fine spectacle, and the Irish in particular caught the eye of onlookers. 'Look at the Irish now down there in the plain', one of them pointed out to another, 'there are two squadrons; and the Earl of Kildare leads the first, and the Earl of Desmond has drawn the second from the wild mountains'. Before the Irish there marched the English and the Scots, and after them there came the Swedes and the Norwegians, together with men from Thule and from 'remote Iceland'. The Christian king of kings must surely have been gratified by so generous a response.

Nevertheless, we cannot help but wonder whether the response was not also meant as a sign of just how much Charlemagne was down on his luck. These men from 'beyond', we are told, are 'natural enemies of peace' and, hailing as they do from caves and woods, 'they have a hairy face, and chest, and ribs, and backs, and arms, and legs, like beasts'. Could any good really come of such men? – we are bound to ask. True, it may well be that 'beyond' started after Ireland, and that the designation 'hairy beast' was therefore not intended to apply to the Irish. The phrasing is ambiguous, and since Ariosto (1474–1533), in whose *Orlando Furioso* all of this takes place, is a master of ambiguity, it is hard to tell (*Orlando Furioso*, let us not forget, is one of the artful works which helped Zanina send Gandino from Bergamo to the 'Kingdom of Cornwall'). At the same time it is difficult not to suspect that Ariosto did indeed have ulterior motives with respect to Ireland, all the more so as the Earls of Kildare and Desmond, who only appeared on the scene after the twelfth-century invasion of Ireland by the Anglo-Normans, are quite out of place in the ninth-century world of Charlemagne.[1]

Admittedly, there is nothing which poetic licence cannot bring about, and Ariosto never shirks from using his to the full. However, his instinct as a Ferrarese was honed in equal measure to his instinct as a poet. He was as proudly Ferrarese as Landino was a Florentine. He may not always have seen eye to eye with his patrons, the arrogant and demanding Cardinal Ippolito d'Este (1479–1520) and the latter's brother, Duke Alfonso I (1476–1505–34), but he loved Ferrara and he served his masters well: as secretary, as court official, as ambassador, as governor of one of the remotest provinces of their duchy, as playwright, as provider of court entertainments, and above all as purveyor of enchanting tales of damsels in distress and knights in shining armour, which, in the first, forty-canto edition, published in 1516, he entitled *Orlando Furioso by Ludovico Ariosto from Ferrara* [*Orlando Furioso de Ludovico Ariosto da Ferrara*], and in the third, forty-six-canto edition, published in 1532, *Orlando Furioso by Master Ludovico Ariosto, Nobleman of Ferrara* [*Orlando Furioso di Messer Ludovico Ariosto nobile ferrarese*].

Orlando Furioso was conceived primarily as entertainment; but it is the work of a learned poet, full of references to other works, and it is also highly charged politically, with many allusions to the events of Ariosto's day. Since one of Ariosto's primary inspirations was Dante's *Comedy*, which he is bound to have read in Landino's edition, and since the rivalry between Ferrara and Florence, the Este and the Medici, was intense at the time, it is easy to imagine that, even as he paid homage to Florence's divine prophet, Ariosto would have wanted to deflate the pretentiousness of those who, standing on his shoulders, tried to pass themselves off, dwarves that they were, as giants too. If, to prove their greatness, the Florentines boasted of the Irish success story of the Gherardini brothers, why not have those same Fitzgeralds – Ariosto could well have asked himself – reduced to being lords of 'hairy beasts' and suggest that they might just as easily be of hindrance as of assistance to the Christian cause? Anyhow, was it not the case that Ireland was the land of make-belief?[2]

The spectator at the review of the troops outside London who is told to have a good look at the Irish is Ruggiero, the mythical ancestor of the Este, whom Ariosto makes into one of the main protagonists of his poem. As errant as Ariosto was domesticated and as libidinous as Ariosto was uxorious, Ruggiero has just completed a tour of the world, on his flying horse, travelling all the way from the Pyrenees to China, westward over the vast emptiness of the ocean (in the ninth century, the New World had not yet been discovered), and from there back to western Europe, via the Himalayas, Siberia, the Caspian Sea, Russia, Prussia, Pomerania, Poland, Hungary and Germany. Nowhere, though, is Ruggiero's curiosity very much aroused. Every bit the pampered tourist, he is less concerned by what he sees than where he stays – 'bad lodgings', we are told, 'he avoids at will' – and only on one occasion does Ariosto spend more than a couple of words to describe a land that falls beneath his (vacant) gaze. Having had his fill of the parade, Ruggiero visits England 'from side to side', and thereupon 'he goes towards Ireland': 'and he saw fabulous Hibernia, where the little old saint had made the cave in which there seems to be such grace that man can purge his every wicked sin'. In the whole wide world, the only place which deserves Ariosto's attention is *Ibernia fabulosa*.[3]

True, Ariosto does not linger there for very much longer than does Aeneas Silvius Piccolomini, and he too seems to have his doubts as to whether the fables are worth remembering. And perhaps, like Fazio, and possibly Silvestri, he also thought that *Ibernia* was but a part of *Irlanda*. Of one thing, though, he was quite sure: his readers were bound to understand him. There was no need to identify the 'little old saint' or to put a name on his purging 'cave'. Or to explain why he himself harboured doubts as to the effectiveness of the purgation. It would have been common knowledge. We too know what he is talking about, for it is that same cave which led Fazio's yearning to get the better of his reason and permitted him to call the Irish 'sweet', even though, in the end, he was too frightened to go inside. We also know of the cave from Silvestri and from Fra Mauro and from Berlinghieri. It is the cave of St Patrick: St Patrick's Purgatory. But why, we may wonder, has St Patrick now become a diminutive saint? And why the scepticism about his Purgatory? Was there not a time when it was held in greater awe?

St Patrick's Purgatory, on Lough Derg in present-day County Donegal, was first brought to the attention of Europe's faithful through the so-called *Treatise on St Patrick's Purgatory* [*Tractatus de Purgatorio Sancti Patricii*], written in 1184 by one Brother H., a Cistercian Anglo-Norman monk from the Abbey of Saltrey in Huntingdonshire, in England, who had himself heard of it from another monk, Gilbert, from the Abbey of Louth in the diocese of Lincoln, also in England, who in 1148 had been sent to found a monastery in Ireland, where he had had as an Irish-speaking guide and interpreter a man named Owein, who claimed to have only just returned from the Purgatory. The *Treatise* became extremely popular throughout the continent and survives in one hundred and fifty manuscripts, as well as many translations into European vernaculars. The Purgatory itself, it is supposed, had come into being for reasons linked to the pagan past of Lough Derg and to its association with a long established Christian anchorite tradition, as well as to more recent conflicts between the newly arrived Cistercians and Augustinians, vying for control of the area, and to struggles among local lords and bishops; but its existence was of far more universal significance. It concerned the Church as a whole, serving to bolster both its authority and its eschatology, and according to one famous theory, the *Treatise* was nothing less than the 'literary birth certificate' of purgatory true and proper.

Of course there was no such thing as purgatory true and proper, and before Dante nobody even knew what it actually looked like or where it really was. There is no mention of purgatory in the Bible, only of purgation, and the dogma was not defined by the Catholic Church until the Second Council of Lyons in 1274. According to that same (controversial) theory, put forward by the renowned French scholar Jacques Le Goff, purgatory was in fact 'born' in the schools of Paris, then the leading intellectual centre of the western world, in the second half of the twelfth century, during a period of 'great intellectual and moral revolution', which saw the strengthening of the notion of personality and personal intentions and responsibility, and greater importance being attached to contrition and confession.[4] In effect purgatory, by positing a distinction between mortal sins and venial sins, and stressing the belief in penitence and penance, offered sinners a second chance of being saved. It was a waiting room, so to speak, where venial sinners and repentant sinners could purge themselves and make reparation for their wickedness, before being allowed to proceed to paradise. But as Dante was to illustrate so memorably, and as many churchmen would have explained in their sermons, purgatory was by no means a bed of roses. The punishment you would be made to endure there would be extremely harsh and long-lasting. As a result there evolved the hope and the belief that in some ways that punishment might be mitigated. Prayers for the dead came to be seen as efficacious, that is, as having the power to move God to release your loved ones from their purgatorial tortures before the end of their sentence; and indulgences, a kind of safe-conduct which guaranteed a swifter exit from purgatory, began to be offered by the Church, very often in return for a monetary reward.

As well as strengthening the unique sacramental nature of the Church – especially *vis-à-vis* the eastern Orthodox Church, heretics, and in due course Protestants, none of whom accepted the belief – the 'birth' of purgatory also helped to consolidate the supremacy of the papal monarchy, which was being engineered at that very time by popes intent on rendering effective the universality of the Church's authority, through a process of bureaucratic and legal reforms, aimed at defining and bolstering its institutional and temporal power, and a process of dogmatic and ecclesial reforms, aimed at proclaiming and releasing the priesthood's special charismata, by virtue of which priests stood apart from and above the laity. Purgatory vested in priests, and ultimately in the pope, the exclusive privilege, herebelow, of sanctioning salvation. What is more, by its identification with the Church Militant rather than the Church Triumphant, that is, with the believers who were still fighting to be saved rather than those who were already saved (in paradise), purgatory effectively extended the Church's and the pope's jurisdiction into the afterlife. All told, purgatory was the proof and the guarantee that the power of the Church was absolute. And the proof of that proof was provided by the 'little old saint' and his 'cave'.

In Italy the story of that cave became famous thanks primarily to *The Golden Legend* [*Legenda aurea*], a collection of saints' lives composed by the Dominican friar turned Archbishop of Genoa, Jacopo da Varagine (c.1228–98). As told by Varagine the story differs little from the account in the *Tractatus*, except that Owein has been renamed Nicholas. It tells of how St Patrick, having tried but failed to convert the Irish, asked God for a sign which would strike terror in their hearts and thus lead them to see the light. God ordered him to draw a circle on the ground, which St Patrick did,

> and lo! the earth within the circle opened up and an enormous and very deep well appeared there, and it was revealed to the blessed Patrick that that was a place of purgation, in which, for those who wished to descend, there would be no further penance left nor would they know any other purgatory for their sins, but that many would not return from there and those who did return would have to stay there from the morning until the following morning. Many therefore went in, who then did not come back.

A good few years after St Patrick's death, the place was visited by a man named Nicholas who, having sinned excessively, wished to do penance. He was allowed in, but warned that he would experience many dangers and temptations, and would inevitably succumb to them if on every occasion he did not cry out 'Jesus Christ, son of the living God, have pity on me who am a sinner'. And so it happened.[5]

He was saved from the blandishments of demons who promised to protect him, from rabid beasts who threatened to devour him, and from countless other tortures. He was also saved after being hurled into a wide gulley that produced thick smoke and a frightful stench, and into which sinners were plummeting as though they were made of lead, yet bouncing back like sparks, only to be repelled by the devils who announced to Nicholas: 'this place you see is hell, where lives our master Beelzebub'. The invocation also saw him safely across the narrowest of bridges, as shiny and slippery as ice, under which, amid a terrifying din, there flowed an enormous river of sulphur and of flames; and on the other side he came to a most pleasant meadow, where all sorts of flowers gave off a wondrous odour and where he was greeted by two handsome youths, who lead him to a splendid city, made of gold and gems, from the gates of which there emanated a scent so marvellous it erased all memory of his sufferings. 'This city', they said to him, 'is paradise', but as he tried to enter, they told him that first he was to go back to his family, along the route he had just taken but where the devils would now flee in fright before him, and only then would he be allowed in. 'And so Nicholas, going up the way he had come down, found himself restored and, telling everybody everything that had happened to him, after thirty days he rested happily in the lord'.[6]

Like other places of pilgrimage, especially the three most famous ones at the time (Jerusalem, Rome and Santiago de Compostela), St Patrick's Purgatory offered pilgrims the promise of a fail-safe salvation and the possibility of gaining an indulgence; but otherwise it was quite unique. Not only did it guarantee that after death they could go straight to paradise, without passing purgatory, but it also afforded them visions of the afterlife. Indeed, according to some interpretations, it literally transported them to the afterlife, allowing them to experience hell, purgatory and paradise for themselves before dying, or in other words, to be suspended between two worlds, this one and the next. However it carried a terrifying warning: 'many went in, who then did not come back!' For sure, visiting that cave was not for the fainthearted, however old or little its patron saint may be. So now we understand why Fazio was so scared. It was not just authorial *Angst* which haunted him. It was the fear of oblivion.

For all we know, there may have been many who succumbed to that fear, and many who went in never to come back. It is said, no doubt rightly, that St Patrick's Purgatory was one of the major European destinations of pilgrimage during the late Middle Ages and early Renaissance and that it attracted pilgrims from all over Europe. We do indeed have testimonies from pilgrims from different parts of Europe, not least Italy, who claimed to have visited it, but all told there are not really that many; and not a few accounts, by people who did or did not go there, display a degree of indifference towards the place. Petrarch for instance, even though he highlights with an asterisk the brief reference to St Patrick's Purgatory in Gerald of Wales, never talks about it, and he certainly does not include it among the features that make of Ireland an almost 'happy' land. But that need not surprise us. Petrarch positively wallowed in the idea of guilt and sinfulness, and he would probably have been amazed by the notion that one should or could want to rid oneself of what was after all the stuff of life (and death). Sin is what we are born to, and without sin Petrarch would have had nothing to say, about Laura, or about himself.

Silvestri too, all told, pays little attention to the Purgatory, and even seems to doubt whether it really exists, or at least to question Bishop Milo's assertion that it is to be found in *Ibernia*, when he himself feels it should be in *Iberia*. More surprisingly, the location of the Purgatory is all the two appear to have discussed on the matter, despite Silvestri's irresistible fascination with Ireland's 'amazing to say' and notwithstanding what presumably should have been Milo's professional obligation to talk up the benefits of Ireland's purgatory – assuming, that is, Milo was indeed the archbishop designate of Armagh. If St Patrick's Purgatory became the 'fabulous' attraction which it did, it was thanks primarily to Milo Sweetman's predecessor as Archbishop of Armagh, Richard FitzRalph, who, especially during the time he spent at the papal court in Avignon, engaged in a sustained publicity campaign to promote the pilgrimage to Ireland. Should Milo not have done the same? – we cannot help but wonder. Unless of course he was another Milo altogether (which, as we noted, is a distinct possibility), or else a figment of Silvestri's imagination. Or perhaps Silvestri's imagination was more akin to the imagination of Fra Mauro, a century or so later. Fra Mauro too, as we have seen, considered Ireland to stand for marvels by antonomasia, yet for him St Patrick's Purgatory was no more than a landmark on a map, just like any other; and that is also how it was to be for Berlinghieri, for all that Berlinghieri took his inspiration from a poet for whom purgatory was an undeniable reality. Nevertheless, the fear which could arise from the thought of gambling this world on the chance of a better deal in the next, by visiting St Patrick's Purgatory, was by no means a thing of the past. Indeed, at about the time Silvestri passed away, the Purgatory was visited by a Florentine merchant, Antonio Mannini, who was so frightened by what he witnessed that all he dwells on in the letter which he sent to his brother back in Florence, in February 1412, to inform him of the pilgrimage, is his exceeding fear.

Mannini had gone to Ireland to do business, but the business had turned sour and he therefore planned to return to Italy. God, though, had had other plans for him. 'Many times', he tells his brother, 'I made ready to start, sometimes there was no ship, sometimes the wind was contrary, and sometimes I was without money', and as he waited for his luck to turn,

> it pleased God that there should come hither a noble gentleman of Hungary, named M. Lorenzo Rattoldi, on his way to the Purgatory of St Patrick. I visited the gentleman, and held long converse with him; and seeing his holy intention and devotion, for he had come from the head of all Christianity in the world to the end of the earth for the good of his soul, and that I was here close to the said Purgatory, and in such pain and trouble, considering what my life had been hitherto, sinful, and thinking but of worldly things, comforted by good hope I turned to God, and inspired by His mercy I resolved and determined for the good of my soul to go to the said Purgatory, hoping and believing in the divine mercy that perhaps in the future I might have better grace and fortune and an end to my long tribulations.[7]

The 'divine mercy', as it turns out, was uncooperative. Although Mannini was convinced the pilgrimage had done him good, so much so that he resolved to stay in Ireland and requested of his brother that he make arrangements for his wife and child to join him, a year later he was back in Florence. However, it could well have been his fear, and his fragile constitution, which were to blame.

To the guardians of the place – the prior and the canons of the priory which stood on an island next to the one on which the Purgatory was situated – that fear and fragility must have been all too evident, for they did their utmost to discourage Mannini from going through with his intentions and in the end would not allow him to spend more than five hours, rather than the usual twenty-four, 'shut up in the Purgatory'. To us it is also evident. Mannini's letter says next to nothing about that country at 'the end of the earth': all it mentions is that 'the Purgatory lies in a lake among high mountains' and that 'the whole of the lake is the coldest country in which I have ever been; the mountains of Brigha [i.e. Brig, in modern-day Switzerland] are not cold in winter compared to that place'. Neither does it say anything about 'the many visions and temptations' Mannini claims to have experienced inside the Purgatory, and it says precious little about the 'four graces' he was 'instructed' to ask of God, except to mention that regarding 'three of these it is best I should be silent; for the fourth, I asked God's mercy that I might recover my honour and state, and that of our house'. Instead, as we said, the letter is all about Mannini's fear, which was such, it would appear, that at one point it made his hair stand 'straight upon end so that [he] could scarcely smooth it down with [his] hands and make it lie flat', at another it almost caused him to drown in the lake, and when at last he stood before the Purgatory, it made his heart 'flutter and beat' uncontrollably, so that even he was forced to wonder 'how I endured it so long'.[8]

So long in fact did the fear endure that over a century later it could still frustrate a pilgrim's best intentions, just as it does our own curiosity. In 1517, twenty-five years after the publication of Berlinghieri's *Geographia* and a few months after the publication of the first edition of *Orlando Furioso*, there went to Ireland that Italian pilgrim from Vicenza who, as we mentioned in the Introduction, is possibly the only Italian Ireland-watcher of the Renaissance actually to have visited the place and who, we will now find out, was also so overcome by fear – or so at least he claims – that he too lost his nerve at the last minute: in other words, Francesco Chiericati (c.1480–1539), the soon-to-be house guest of the Countess of Pandino and the friend of that other noblewoman with imperious eyes, Isabella d'Este (1474–1539), the Marchioness of Mantua and a sister of Ariosto's patrons. Isabella had a keen interest in foreign lands and foreign travel but could only ever travel by proxy. As soon as she heard that Chiericati was to be sent to London as papal nuncio therefore, she charged him with going to Hibernia and finding out the truth 'regarding the fables that are said about the island and St Patrick's Well'. And Chiericati duly did as he was told. Taking advantage of an outbreak of the plague in London, he travelled to Dublin, and from there to Lough Derg and back, and when he returned, as instructed, he sent a letter to Isabella, in which he regaled her with his own version of the 'fables', even though, he added, 'they are of little account'. However he trusted that her 'exceptional mind, which does not only want to learn of great things, but also of the smallest', would allow her to take 'everything in the spirit in which I write it'.[9]

The spirit, which seems tailor-made to satisfy the most demanding *Pandinesco*, is one of condescending amusement, mingled with a degree of annoyance at having to forego his creature comforts in order to satisfy his friend's 'exceptional mind', together with a touch of amazement, not least at the size of the salmon he was allowed to fish in Downpatrick, courtesy of a one-hundred-and-fourteen-year old Italian bishop, 'which in Italy would be worth a lot and would be greatly prized', at the quality of the 'Scottish' pearls produced by Ireland's many rivers, of which 'the peasants find so many it is quite amazing, with which they do a very good trade', at Ireland's climate, 'extremely temperate and warmer than it is in England, which is quite amazing', at the 'great faith' of the Irish and their devotion to the pope, who 'they say is their king', at their 'different lords, who are little more honourable than our peasants', at their being 'worth a lot in arms, because they are always making war against each other', and at their total disregard for private property, which they justify on the grounds that 'it is we who are the beasts, since we make our property of the goods of fortune, and that they themselves live naturally, because everything must be in common'.[10]

Besides Downpatrick, thanks to its Italian bishop, and up to a point Armagh, thanks to 'an abbey of canons regular', the only place Chiericati found remotely tolerable was 'the city of Hibernia called Dublin', where he was 'honourably entertained by the most reverend archbishop and by the illustrious Earl of Kildare, who is the viceroy of that island [...] is a man of substance and has all the civility of England' (being from Vicenza and connected to the ruling family of Ferrara, Chiericati could obviously not claim that Kildare had all the civility of Florence). After that things got progressively worse, as Chiericati and his party came across more and more 'beast-like people' and ruined cities. Nevertheless they kept going, passing through a landscape which, though initially 'very pleasant by the sea', became increasingly 'desolate', with its 'thieves and woods, lakes and bogs', covering the eighty odd miles to St Patrick's Purgatory on foot, crossing over to the island on which it is situated in 'a dugout beech trunk', after having paid 'a penny a head', enduring the 'nine days and nine nights' of rituals associated with the pilgrimage, and to cap it all suffering 'the greatest penance' of seeing most of their provisions run out.[11] Yet after all that, when it came to entering the cave, Chiericati got cold feet:

> I was not able to see the place because I did not want to look inside, frightened by the things which are said; but I kept three strides away from the door. And the canons entered with two pine torches, and I looked straight ahead, at everything, but you see nothing else but rock.[12]

As it so happens, those who did go in did not fare very much better:

> Of those who went in when I was there, two saw such apparitions, one of whom came out almost senseless and, being asked many questions, said he was beaten extremely violently but that he didn't know by whom; the other said that several very shapely women appeared to him, who invited him to eat with them and enjoy himself, offering him fruit and many kinds of dishes; and he says that he almost let himself be tempted, because of the great weakness he was in. The others said they saw nothing at all, nor felt anything, except for the cold, hunger and a great weakness.[13]

Clearly, the 'little old saint' had ceased to work his wonders. Or else it took an altogether different class of pilgrim to draw upon himself the full 'grace' of his purgatorial 'cave', a pilgrim who was not so faint of heart.

There had been a pilgrim of the sort in 1358, or so at least the records say. His name was Malatesta Ungaro (Malatesta the Hungarian: 1327–72), he belonged to the ruling family of the city-state of Rimini, and he was nicknamed 'the Hungarian' because in 1347 he had been knighted, for services rendered on the battlefield, by the king of Hungary. He was the brother of a close friend of Petrarch's, Pandolfo Malatesta, and in the latter part of the fifteenth century he was made into a protagonist of what was then Petrarch's most popular work. Nowadays Petrarch is known primarily for his collection of love poetry, the *Canzoniere*, and hardly anyone ever reads his bombastic *Triumphs* [*Trionfi*] anymore, but in the early Renaissance it was the other way round; before becoming famous for the *Canzoniere*, Petrarch was revered as the author of the *Trionfi*, the *terza rima* allegorical poem which sang of life's triumphers. The early Renaissance was triumphalist, it approved of moralizing, it had a fondness for Dante's *terza rima*, but above all it welcomed works which satisfied its thirst for fame and greatness, as well as its desire to find out more about the past. The *Canzoniere* focuses on the private torments of a soul which seeks its own annihilation or redemption. The *Trionfi* is like a roll-call of victorious worthies from the distant, and not so distant past, who parade before the eyes of an inquisitive narrator.

Among the many who march past in Love's cortège, already gone before you even notice them, is an anonymous 'couple from Rimini', and according to the author of a commented edition of the *Trionfi* published in Parma in

1473, that couple is 'Viola Novella, the wife of someone called Coça Battaglia', and her lover 'Malatesta, lord of Rimini' who, when Viola had been killed by her jealous husband, 'out of love for her went to the Purgatory of St Patrick'. Nowadays we would most likely presume the couple to be Dante's star-crossed lovers Paolo and Francesca da Rimini, and so in fact did many readers of the time: in the far more popular commentary on the *Trionfi* by the Sienese physician and philosopher Bernardo Ilicino (1435–76), first published in Bologna in 1475 and republished a further twenty-three times over the following fifty years, that is precisely who the 'couple from Rimini' are said to be.[14] Nevertheless the credentials of Malatesta Ungaro as a rival to Dante's Paolo were well established, and so famous had he apparently become for his daring pilgrimage to Ireland that in Italy he was made to rank third, after St Patrick and Nicholas, of those who 'went in' and subsequently did 'come back'. Silvestri, when discussing the well-like hole that is to be found in *Iberia* – or *Ibernia* – next to the monastery of the blessed Patrick and which leads to the underworld, claims that only two people are known to have returned, Patrick himself and Nicholas, but then specifies: 'although our own age has added a third, namely the noble Malatesta knight from Rimini, surnamed the Hungarian'. The claim is then repeated by Bandini, in his *Fons memorabilium universi*. Bandini, though, appears somewhat more sceptical. 'It is written', he says, 'that only Patrick and Nicholas returned, although in our own age the bold knight Malatesta of Rimini, surnamed the Hungarian, has boasted publicly that he himself returned from the said well'. Should we believe everything that is written – Bandini seems to query – especially about braggarts? He may have had a point.[15]

The closest piece of evidence we have regarding Malatesta's pilgrimage to Ireland is a letter of congratulations sent to him by a friend of his, one Checco Rossi da Meleto, and all we really know about Checco is that he was also a friend of Petrarch's and Boccaccio's, with whom he had an exchange of poems about the freedom of the will and the way God punishes us for our wickedness by inflicting lightning and other natural disasters on us, poems, that is, about the very phenomenon the absence of which, for Petrarch, made Ireland into a go-there happy kind of place. But it is in fact a very different kind of place which Checco writes about. Checco's is an Ireland fit for a hero – a new Aeneas, Theseus, Jason, Hercules, Julius Caesar, or Alexander the Great, we are told – and an Ireland unsuited for anyone who puts 'pitiful *otium*' before 'glorious deeds'. It is in the 'far west', where 'only the occasional shepherd dwells' and where woods signify not peace but terror: a 'wilderness of wooded mountains', a 'barren land quite devoid of any charms but with terrifying beasts aplenty' and whose 'awesome aspect fills human eyes with fear'. However, Malatesta Ungaro was not a man to be put out by such a prospect. 'Unbeaten by travails' or by 'hardships which a nobleman brought up in comfort might have found too much to bear', he succumbed neither to 'the lack of food', nor to 'the lack of sleep', nor to 'the cold', nor to 'thirst, however many times he had to endure them all, as he trudged through the wastelands of Hibernia'. And in the end he was duly rewarded.

His pilgrimage had been worth every bit of the hardship and he returned to Italy in glory. He was

> seen by England, now blessed in its king and waxing with its lucky star, by the last frontiers of Britain and by furthest Thule, as well as by the mighty Atlas, where the sun goes down, and by almost all the western reaches of the world. Yet that is nothing! For then, of [his] own free will, [he was] welcomed by that blackest of hollow places, hidden among caverns, where a night of dense fog and the shadows of darkness conceal the gloomy entrance to the underworld. And yet, intrepid, [he] came back from there, whence very few who enter do return, such are the errors – I declare – that draw mortals to the blind ways and the woes of secret places.

There is no doubt: what Malatesta had visited was no mere cave dug by a little old saint. It was nothing less than a 'voracious chasm' which 'fills human ears with fear'.

What exactly the 'bold knight' actually saw or did in that chasm, we are never told. Neither are we specifically informed why he went there in the first place. Checco does mention that it was because of 'that uncommon love' of his, but since he makes no reference to Viola Novella, or to any other woman for that matter, we cannot help but suspect, given the tone of the letter, that it was the love of adventure rather than romantic love which drew Malatesta to Ireland. We are even bound to wonder whether he went there at all. The letter is so exaggerated, so full of hyperbole ('furthest Thule' indeed!), that it sounds more like a joke among friends, or even possibly like an act of revenge on the part of a disappointed suitor. Just as Ariosto was to present *Orlando Furioso*, in which he often pokes merciless fun at Ippolito d'Este and his family, as the gift of a 'humble servant' in need of protection and reward, so Checco signs the *Letter to the Magnificent Lord Malatesta the Hungarian of the Malatestas in which He is much Praised especially for His Journey to the Well of St Patrick* as 'the least of his clients', praying that he should 'not show disdain for the menial labour of an impoverished servant'.[16] So might it be that Bandini was right? Might it be that Malatesta the Hungarian, far from being a hero, was simply a braggart who boasted of his pilgrimage to Ireland without ever having been there?

What is for sure is that he acquired quite a reputation as a fake pilgrim. In the tenth of the *Three Hundred Tales* [*Trecentonovelle*] by the Florentine merchant turned politician and writer Franco Sacchetti (c.1333–1400) we witness Malatesta taking part in a pilgrimage to the Holy Land, probably in 1349, in the company of his uncle, Galeotto Malatesta, Lord of Rimini, and of the then famous court jester and musician, Dolcibene dei Tori, who is the protagonist of the tale. The pilgrimage was to inspire Dolcibene to compose 'an oration to Our Lady, to thank her, telling of all the holy places he had visited overseas', but as described by Sacchetti there was little about it that was holy. Its highlight is an incident in the Valley of Jehoshaphat, where Dolcibene, to the great amusement of his companions, rushes to lower his breeches and 'release his body's necessities', so that, come the Day of Judgement, he would 'find the place' again and not 'drown in the throng'. Admittedly all Malatesta Ungaro does is watch and laugh, but on considering the story, the mind of the reader is inevitably drawn to Ireland's 'voracious chasm' and made to wonder what on earth the 'magnanimous hero' may have been up to there, and why his encomiast writes, as he says he does, with a 'trembling hand', a 'faint heart', and a mind that 'vacillates as it readies to sing his praises'.[17] The reader is all the more bemused as Chiericati too, writing about his own pilgrimage to Ireland, was to highlight the link between scatology and eschatology. 'On one side', says Chiericati, referring to the pilgrims who 'went in',

> the rock is pierced, and through a little hole a vase is passed and taken out again for their necessities. And always one of the canons stands by the hole, preaching that they be steadfast and that they do not let themselves be overcome by diabolic temptations, because, they say, very many different apparitions take place there, and many of them, who they say gave into temptation, come out oblivious and mad.[18]

If it was not Checco who was a liar, was it Malatesta Ungaro who was 'oblivious'? Or 'mad'?

He himself, it would seem, feared that that is what others might one day believe. He therefore sought to gather as many proofs as possible of his uncommon heroism, by ensuring that his pilgrimage was certified in writing by the prior in charge of St Patrick's Purgatory, by the king of England's justiciar in Ireland, Amaury de Saint Armand, and finally, on 24 October 1358, by the 'blessed' king himself, Edward III (1312–27–77). If all these proofs do indeed survive, as they would appear to do, who are we not to believe what they say? But as it turns out we have also read Chiericati and we know only too well of the 'spirit' in which pilgrimages could be carried out. At the end of the account of his pilgrimage, Chiericati explains to Isabella d'Este that there is a guest-book, as it were, in which the pilgrims must all be recorded:

> The following day they came out half dead and they were restored as well as possible, and they were entered into the book, which is in the church, in which all those who go in are listed. The first person listed was Guarino of Durazzo, which I thought was a fable. But now I have seen that he was indeed listed there, a long time ago, in a parchment book.[19]

Normally, for a humanist such as Chericati, if it said so in a book, then it had to be true. On this occasion, though, he does not say so in so many words. Guarino's name was listed in the book all right, but Chiericati seems to be unsure as to whether or not to believe it. And as it happens – and as Isabella was bound to know – he had every reason to be suspicious, for Guarino was in fact a fictional character, born of the imagination of one of the most engaging writers of the Italian Renaissance. Nevertheless, for all that he was 'fabulous', it was Guarino, it may be safely vouched, who did most in Italy to turn St Patrick's Well – *il pozzo di san Patrizio* – into a household name.

Guarino of Durazzo, otherwise known as Guerrino il Meschino [Poor Little Guerrino], is the brainchild of Andrea Mangiabotti, or Andrea da Barberino, as he is more commonly referred to, the Florentine storyteller who authored the *Royals of France* [*Reali di Francia*] mentioned in Chapter One, in which 'England and Ireland had become Christian' while 'all the other provinces were Saracen and pagan'. Exceptionally for the time, Andrea was able to live by his pen, it would seem. His life spanned the years between the revolt of the Ciompi (the 1378 uprising of the *Lumpenproletariat* that sent shockwaves through the closed ranks of the guildsmen who had hitherto monopolized the political machine of the city-state) and the rise to power of the ruthless Medici (the bankers who managed to oil the machine so that it would run to the tune of their own ambitions and desires), and he was a professional storyteller, who not only wrote his stories down but also recited them in the squares of Florence for the entertainment of the populace, and whose fans included not only the city's lower orders but also members of its most prestigious families. Nowadays his works are all but forgotten, except by a few specialists, and even they, at times, seem to believe that it is best they be forgotten; yet it would be no exaggeration to say that few other Renaissance authors enjoyed the enduring popularity which Andrea did, especially thanks to Poor Little Guerrino.

Guerrino il Meschino, of which there are eleven known manuscripts, was printed no fewer than fifteen times before 1500, another fifteen times in the course of the sixteenth century, six in the seventeenth, nine in the eighteenth, and more than thirty in the nineteenth century, when it became, often in abridged or popularized versions, the favoured reading of the more humble classes, from north to south of Italy, but above all in Sicily. There are also reports that it accompanied many an emigrant across the Atlantic and that it was read and passed around with eagerness in Italian neighbourhoods in the cities of the USA. What must have made the work so popular is not only the fast-moving pace of its plot and the adventurousness of its characters, but also the author's quite unique sense of humour. Its eponymous hero is a foundling from Durazzo in Albania (modern-day Durrës), who grows up in Constantinople, becoming best friends with the son of the emperor, until he discovers that he is not who he thinks he is, and so embarks on a ten-year search for his parents, which will take him to every corner of the world, and the underworld, until at last all is revealed to him, or almost all, at St Patrick's Purgatory.[20]

Guerrino lived at the time of Charlemagne, but it is a time which poetic licence has conflated with the eleventh- and twelfth-century Norman settlement of southern Italy, and which the author's perceptiveness has stretched to include the threat posed in his own day by the advance of the Turks. Guerrino's family, in a sense, were relatives *avant la lettre* of the Earls of Desmond and Kildare. They issued from 'the race of Burgundy' (*alias* Normandy) which, on Charlemagne's orders, had been sent to Apulia and Calabria to repel the 'African' invaders, and Guerrino's uncle, Guiscard, was invested with the kingdom of Apulia, while his father (the namesake of the Irish bishop who even as Andrea was labouring to produce *Guerrino* disagreed with Silvestri about the location of St Patrick's Purgatory) was created Prince of Taranto (the adoptive homeland of St Cathal), which he used as a springboard to seize Durazzo from the Turks. There he married the sister of the former rulers of the city and fathered Guerrino but was soon defeated by his brothers-in-law, who threw him into jail, where he remained, together with his wife, for the next thirty years, until their son had reaped the benefit of St Patrick's grace. Meanwhile poor little Guerrino was bundled on a ship, seized by pirates, sold at the slave market of Salonica, brought to Constantinople, made to believe he was the son of the Greek who purchased him, given a princely education, raised to the status of 'hero of Constantinople' for having saved the city from the Turks, and all of this before he reached the age of twenty, when he found out that his true identity had been kept from him and so his whole world fell apart. Yet right away he set about rebuilding it, and though it took him ten full years to do so, hardly ever did he allow himself to be downcast.

Guerrino's natural playfulness and resourcefulness saw him safely from one end of the world to the other, starting in India, at the trees of the sun and the moon, the oracle which according to legend had once been visited by Alexander the Great, and though he never lost sight of his quest, he nevertheless enjoyed every minute of his travels. Part explorer, part knight errant, part pilgrim, part tourist and part crusader, unlike Ruggiero he travelled with an open mind and an enquiring gaze and was equally at home among Christians, Muslims or pagans, as well as with the denizens of the underworld in the cave of the Cumean Sybil. Indeed, to such an extent does he become a man of the world, as it were, that for most of the novel it is not clear to the reader, and neither to Guerrino himself possibly, what side he will finally embrace. In the end, having seen at St Patrick's Purgatory how in the lowest part of hell, forever damned, there stood not Lucifer, as Dante had imagined, but the Prophet Mohammed, he became (after treating himself to a bit more sight-seeing) the single-minded Christian warrior which his birth and his double baptism (once as a Catholic, once as a Greek Orthodox) had appointed him to be; and admittedly, even before then his instinct had been to fight for the Christian cause, as when he saved Constantinople from the Turks. Yet his instinct did not stop him, when necessary, from fighting for the Muslims, as he did when he lead an anti-crusade which delivered Syria and Jerusalem into the hands of the Sultan of Babylon, or from saving Muslim damsels in distress, as he did when he went to the rescue of Antinisca, the daughter of the king of Persepolis, with whom he fell in love and whom he then married, even before asking her to become a Christian.

No doubt he was trying to hedge his bets, until he was quite sure as to which religion offered the best rewards or which God was more likely to satisfy his yearning. When he was in Jerusalem.

> he secretly spent a night on his knees beside the holy sepulchre of Master Jesus Christ and prayed him for the mercy of knowing his father and his birth, swearing upon the holy tomb to make good his promise to Antinisca and that he would make her a Christian if he found his father.[21]

Earlier he had visited the resting place of Mohammed, where he had turned not his face, as was required, but his rear end towards the tomb, while praying *sottovoce* 'Oh! cursed cause of scandal, may divine justice give you just rewards for the souls that you mislead and have misled through your operations'; but when he was caught by a lynch mob and brought before a judge, he managed to convince the judge that he had done so because he felt too great a sinner to dare look Mohammed in the face, and the judge had thus released him, proclaiming him to be a saintly model for all good Muslims to revere.[22] Guerrino had been grateful for the honour. And how wise he was to keep a foot in either camp was then revealed to him when finally he went to see the pope in Rome, after consorting with the devil in the Sybil's cave. Neither the Sybil nor the devil had been able to tell him very much about his parents, but he was now so totally beyond redemption, having dared to dabble in the occult, that only the pope, he had been warned, would be able to set him on the straight and narrow. The pope, however, was hardly even master in his own home and he did not seem to be too sure himself just why the penance he imposed on Guerrino – visiting St Patrick's Purgatory – would be the answer to his problems.

The papal curia was a far more hostile environment than any Guerrino had encountered so far, and it required all Guerrino's native cunning to get beyond the mocking curialists whose only interest lay in the money in his pocket. So as to watch and learn, he decided to lie low a while, but when one day a group of ambassadors was led into the presence of the pope he seized the opportunity. Hiding in among them, he got as far as the threshold of the pope's inner chamber, and when he was found out and his way was barred yet again, he kicked up such a fuss, ranting and raving at the top of his voice, that he could no longer be ignored. The courtiers would have expelled him on the spot, but the pope's curiosity had been aroused, and while the others laughed and scoffed, he listened carefully to Guerrino's tale, and recognizing him as the hero of Constantinople, he decided that his cry for help was genuine. Therefore,

> he imposed on him a penance, since he had dared, against the command of God's law, to enter where the Sybil was and to go and visit the idols, that is, the trees of the sun, and so he willed that upon his own command he go to the Purgatory of St Patrick, which is under the archbishop of Hibernia [*Ibernia*] on the island called Ireland [*Irlanda*], and first that he go to Master St James of Galicia and to St Mary of Finisterre; and Guerrino thanked him very much when he told him that, having done this penance, you are absolved by God and by me. And the Holy Father prayed him, saying: if you enter into the Purgatory of St Patrick, if you can, come back to me to tell me what you found there: but this I do not command you, but leave it to your will. And Guerrino answered: Oh Holy Father, I will do anything, if I live long enough to get there, but there is only one thing which bothers me, and that is that I do not have money for the journey; and the pope had two hundred pounds given to him […] and this was Pope Benedict III, in the year of our Lord Jesus Christ 824, and the emperor was Charlemagne the Older.[23]

In 824 of course purgatory was not yet 'born' and the pilgrimage to Ireland not yet invented – and, for that matter, Charlemagne was already dead (he died in 814) and Benedict III not yet pope (he reigned from 855 to 858). Still, Guerrino took the money and did as he was told; and poetic licence saw to the rest.

Although Guerrino was now entrusted with a dual mission – to find his parents *and* to provide the pope with a report on St Patrick's Purgatory – he nevertheless took his time over it. First he had to go to Santiago de Compostela and Santa Maria de Finisterra, and on top of that there were plenty more sights to see and adventures to attend to. In any case his parents had been waiting for so long, that a few more months would not make any difference, and the pope, for all that he was curious about St Patrick's Purgatory, did not in truth seem to be so very anxious to know more about it. Had he not said to Guerrino: 'this I do not command you, but leave it to your will'? So why forsake what was, after all, his only chance of growing up? In the inn, deep in the forests of the Apennines, where he had spent the night before venturing into the realm of the Sybil, he had had a conversation with the inn-keeper about the merits of travel and discovery. The inn-keeper had wanted to know where he was from, and Guerrino had replied:

> I come from the whole world and I do not know whence I come nor where I go. [...] Have you ever searched the world? The inn-keeper answered: I have been to Syria, to Romania, to the west, to Spain, to England and to Flanders, and now I have returned to my homeland; I have experienced good and bad, and if I ever have grown-up sons who can earn their living, I will make them search the world, for whoever has not searched the world is not a man.[24]

Searching the world was what this life was all about, just as much as searching for one's parents or oneself, and there was no point in gaining an identity if one did not also gain one's manhood.

So important was this search perceived to be that very soon Andrea's masterpiece was advertised not only as a book of adventures but also as a guide-book to the many regions of the world and their marvels – not least St Patrick's Purgatory. 'In this book', the title of the 1534 edition reads,

> is related a brief history of King Charles the emperor. Then the birth of that most famous knight named Guerrino. And the great battles that he fought with the Turks and the Saracens and with many other sorts of people. And the different animals he found going to the trees of the sun. And the dwelling of the Sybil and her deceits. It also tells of the pains of the Purgatory of St Patrick. From all of which one may understand the situation of all the regions and the cities, and the variety of men and their customs.[25]

Knights errant, in other words, and those who fed on their adventures, could no longer simply err for the sake of it. All the erring had to have a goal, which was to 'understand the situation' of the world, and the otherworld as well. It was a goal in which were mingled the religious zeal of missionaries, the conquering zeal of warriors, the inquiring zeal of discoverers, and also the acquisitive zeal of traders – the traders who ruled Florence and for whose entertainment *Guerrino* was written, and the traders among whom Guerrino himself had grown up. His foster father was a merchant who had made sure he was equipped with the tools which would see him safely and profitably across the world. In particular 'he made him learn languages for the utility of trade, and for sailing he learnt Turkish'. Better still, he ensured Guerrino mastered Latin, and nowhere was Guerrino to reap the benefit of his linguistic achievements more fully than in Ireland, where, we are told, 'he would not have been understood if he had not had good Latin and good grammar'.[26]

Ireland also brought out in him what must have been another of the basics of his education, just as it was of the mercantile and humanist education of Andrea's day, namely the ability to wield the pen. Like the merchants and the humanists, Guerrino felt that everything must be recorded; and he was also moved to do so, it would seem, by a fear not unlike Silvestri's that the 'amazing to say' might soon be obliterated by the new canon of the 'worth remembering'. Thus, having come back from St Patrick's Purgatory and made his way to England, and even though the pope and his own parents were in need of his assistance, Guerrino did not hesitate for a moment when the king of England's generous hospitality offered him the chance to 'have all his journey written down, from the day he was made a slave in Constantinople until his departure from England'. He stayed in England three full months, and later, when his quest was at an end but the Infidel still remained to be defeated, his first impulse, once again, was to commit his news to paper:

> Meschino right away wrote to Constantinople, to Alexander, and he wrote to Persia, to the city of Persepolis, to the beautiful Antinisca, letting her secretly know that he was alive and had found his father; and he wrote to Babylonia, and to Barbary, to confirm he had been to all the parts he said. Also he wrote to Morea and to England,

> to Master Dinoino, and in no time his words were shown to be true, that he had searched the whole world to find out about his birth.[27]

No wonder Chiericati was to feel he had to trust what it said in the book at St Patrick's Purgatory. Not only had he himself been brought up to believe that books always spoke the truth, but Guerrino, with Andrea's help, had clearly been successful at blowing his own trumpet. 'His words were shown to be true'.

True they may have been – or intended to have been – but it was nevertheless a strange kind of world they portrayed; and stranger than most was their portrayal of Ireland. Pilgrims using *Guerrino il Meschino* to find their way to St Patrick's Purgatory would most likely have felt disorientated. Contrary to what Fazio may have led them to expect, they would not have found the crossing – in this case from north Wales [*Norgales*] – to be 'risky', and they would have reached *Sconfarda* [Strangford?] without encountering either 'rocks' or 'storms'; but like Malatesta, they would have landed on an island which was covered in woods and largely uninhabited. It was 'a very temperate region', especially in the south, 'towards Spain', and it did have 'many castles' and 'farms' and 'many heads of cattle', but the reason it had so few inhabitants was that, as an island, it was repugnant to human nature, for human beings, in Andrea's estimation, always prefer *terra ferma*:

> this is the result of human nature which, by habit, partakes more of earth than it does of the other elements, and even though in this region one lives a lot, nevertheless humans live more on the *terra ferma*, because earth proves to be a more natural element and a mother to us.[28]

The kind of nature which Andrea found more congenial was clearly of the type which was mother to city-dwellers like the Florentines, and this must surely mean that, to him at least, it would have been most unlikely that scions of the Gherardini could have settled in Ireland and their descendants taken pride in their metamorphosis into Irish lords. What is for sure is that there is no 'Gherardine empire' in Andrea's Ireland.

There are, however, a number of cities, some of which are even recognizable to modern ears, for all that their location, in modern terms, can at times be rather puzzling. Coming away from *Sconfarda*, 'towards

England', Guerrino and his Welsh companion Dinoino – a long-lost friend he had saved in Africa and met again by chance in *Norgales* – encounter *Docidia* [Dundalk? Drogheda?]. In the south, in a region called *Lancia* [Leinster?], there is *Laumerichie* [Limerick] and *Garefonda* [Waterford?], and one hundred miles from *Sconfarda* there is *Daus*, on the river *Irbausi*, while a further hundred miles separate the *Irbausi* from the city of *Venech* [Armagh?]. More puzzling than most, though, is St Patrick's Purgatory. No longer – or not yet? – to be found on Lough Derg, it is situated on an island in the sea, near 'the last city in the world' and off the coast of a region which, as though in confirmation of Fazio's and Silvestri's hunch that Ireland and Hibernia are two separate places, is called *Ibernia*, unlike the country as a whole, which is named *Irlanda*. Still more surprising, possibly, is the fact that Andrea's Ireland is fashioned according to geographical notions which are the exact opposite of what was to become the standard colonialist geography of the island and which already then was well on the way to becoming established, even among Italians who found the lure of Ireland irresistible. Gerald of Wales's geography was a geography made to measure for empire builders; Petrarch was a strong believer in empires; Fazio, for all that he found the Irish 'sweet', sided openly with the king of England, Edward III, whom he praises for having 'changed the habits and the customs' of the 'harsh and cruel' Scots, by 'whipping them high above the buttocks'; Checco also took the side of Edward, whose rule he declares to be 'blessed'; and Chiericati, entering fully into the 'spirit' of the times, would approve of Ireland only insofar as it is inhabited by men who have 'all the civility of England'. Not so Andrea.[29]

Far from depicting an Ireland which becomes progressively less civilized and attractive the further away one moves from the pale of settlement towards the 'hell' of Connaught and which is increasingly inhabited by thieves, he has the best part of the island lying in the west, and without a thief in sight. It is a region, we are told, which has

> a most beautiful nature, cities that are large and well inhabited, and a good fifty miles of most beautiful countryside, rich in everything, and it is a region that is most strong, both on account of its land and of its people and beautiful fortresses, and they always stand watchful guard at their doors against pirates.

So wondrous is the region that its priests are exempt from the normal rules of their order, truly making of *Ibernia* the land of the blessed: 'there rules over this region the Archbishop of Hibernia, and both priests and laymen can marry, and blessed are those who can tie the knot with them'.[30]

Perhaps Andrea meant this tongue-in-cheek, but there can be no doubt that when he comes to describe the 'holy island' which lies at the western-most tip of that western-most region and on which St Patrick's Purgatory is to be found, the tongue has been totally retracted, to the point indeed where the island is made to surpass even *terra-ferma* Florence:

> which island was then called the holy island, because no person ever dies there, so temperate is the air; and on it there is nothing venomous, neither wolves nor anything worse; and this island, according to the ancients, is called the island of succour; no foxes, nor weasels, nor many other animals that are on earth can live there, and the air is so well tempered there and so propitious that men and women reach such a ripe old age that they grow bored with life, and when they age like that, they confess and prepare their souls and have themselves carried across this narrow stretch of sea that is between Ireland and this island, and as soon as they are in Ireland they die. And this way of doing things is allowed to them by their priests. This island anciently was called the Holy Island, and presently the Isle of Flesh, which name was given to it by seamen, as though they wanted to say that they were the murderers of their own flesh, because they carry their old folk to their death. This people live a holy life and they are friends of God and very religious. And there are no thieves there, nor as many wicked enemies of God and of the saints as there are among us, and since seamen have given them this name, for God's sake tell me, you, what there is today in the world which Christians have not destroyed. The world is full of every type of theft and of treachery and of fraud. What father, what brother, what priest can today trust another? What religion, what love, what faith, what charity? None. All men hate the good and like the evil. The city of Florence for sure is not that holy island, but on the contrary it certainly has the habit of dying, though not the providence to die; and for that reason I regret that nobody can live on that island.[31]

For sure, there is confusion in the passage. It may be due in part to what Fra Mauro was to call 'the errors of librarians'. Or it may be due to a hasty reading of the sources: the ancients, Gerald of Wales, or possibly Ranulf Higden, the *Golden Legend*, Fazio degli Uberti, probably Mannini's letter and also the so-called *Visions of George* [*Visiones Georgii*], the account of the visit to St Patrick's Purgatory, in 1353, of a (real) Hungarian knight from Italy, George Grissaphan. It could also be due to misunderstandings on the part of the 'seamen' whom Andrea may have talked to: what they, and he, call the 'Isle of Flesh' [*Isola Carnara*] is unlikely to have derived its name from the Italian word for flesh [*carne*], but rather from *Kernagh* which, according to the account of the pilgrimage of Antonio Mannini's companion, the Hungarian 'Lorenzo Rattoldi' (i.e. Laurence of Pászthó), is said to be the Irish name of the largest and gloomiest of the two parts into which the island of St Patrick's Purgatory is divided.[32] Yet, for all that he may have been confused about the place, Andrea, like many others, had clearly fallen under its spell and, like Silvestri and Fra Mauro, he had also come to see it as unique. Just as Silvestri and Fra Mauro appointed Ireland as the place, and the only place, to raise the question of the wondrous and its relation to reality, so Andrea chose it as the only place to raise the question of the link between the real and the ideal, or in other words of the most fundamental issue facing any author who sets out to distil the facts of life in the alembic of fiction. Nowhere else in *Guerrino il Meschino* does Andrea bare his soul in quite this manner, nowhere else are his fellow citizens rebuked, and nowhere else is there mention of a place where it would be preferable to live. For Andrea too, there is no doubt, Ireland was a go-there kind of place.

If Guerrino went there, it was primarily to find his parents. Yet having got there, he almost totally forgot about them, so enthralled was he by what he saw, even though at first it seemed that, at his appointed destination, he might not see anything at all. So risky was the undertaking that the Archbishop of Hibernia and 'the prior of the place' did all they could to dissuade him from persevering with his mission. But Guerrino, unlike Fazio, was not one to be frightened off by danger; after all, he had already survived an encounter with the devil. And so, after nine full days and nights of fasting and of praying he finally got his way, and the Purgatory was opened up for him. What he saw inside, which takes up twenty chapters of the book, is a mix of what the *Golden Legend* and the *Visions of George* describe, combined with the then even better known description of the otherworld provided in the *Comedy* of Dante, all coloured by Andrea's own inventiveness. Guerrino's purge, in fact, is more an otherworldly journey in the mould of Dante than a simple dip into the well, so to speak, although by comparison with the Dantean

Comedy the sequence of events has been altered and the typology of sins in part recast. Guerrino goes to purgatory before he goes to hell, and while in purgatory he witnesses how souls are cleansed of sloth, envy, pride, gluttony, lust, vainglory, greed and wrath, in hell he sees the punishment of traitors, adulators and penny-pinchers, of friars, monks and ruffians, of fraudsters, soldiers and sodomites, of the vainglorious, the desperately lustful and the proud, of 'those who in the world made themselves be worshipped', those who were full of envy, and finally of Mohammed and his followers. Thereupon, with his own purgation all but done, he is led across a straight and narrow bridge, towards a garden of delights, from where, through a door that stands ajar, he is afforded a glimpse of the earthly paradise. Such is the beauty of the place, that he would happily opt there and then to head straight for the afterlife, but he is not allowed to. Before he joins the blessed, he is told, he must return to earth and 'make known to the incredulous the Purgatory of Saint Patrick, and the pains of hell, and the other mysteries which you have seen and will see'.[33]

At every step of his journey, like Dante, Guerrino observes with awe and fascination, and everywhere he asks questions of his guides, who among other things tell him of the origin of St Patrick's Purgatory, explaining (almost in defiance of what it says in *The Royals of France*) that in Ireland 'the early Christians were not people who cared to live in any way but in vanity and they did not attend to the growing of the Catholic faith, and so did not make very much of an effort to have anyone baptized', wherefore St Patrick, 'being religious and a friend of God', requested and received from his Lord the mercy of converting them to Christianity.[34] Not even when he is about to leave the place does Guerrino desist from asking questions, and almost on the threshold of the Purgatory he still wants to be told about the nature of the angels. And then, almost as an afterthought, he remembers to enquire about his parents. For his guides, the angels cause no problem, but on the subject of his parents they are tight-lipped. All they will confirm is that his parents are alive and of royal blood. They are prepared, however, to let him catch a glimpse of them, and before he has the time to realize what is happening,

> through a door there came two people, a man and a woman; the man had much white hair and a big, thick, wavy beard, and all his clothes were rags; the woman

> was hairy all over, had bare feet, long nails, and scabby legs which in several parts revealed raw flesh; and both the man and the woman were like that. Those saints said to him: have a good look at them, for you will find your father and your mother in this very shape; and he looked at them from in front and from behind, so as to recognize them, and he asked them what their name was and they did not answer.[35]

Having clinically inspected every nook and cranny of the *Doppelgängers*, Guerrino then departs without a word. The royal preview had clearly been a disappointment. Kings and queens, he must have felt, are surely meant to look more regal. All the same the cave of the little old saint had worked its magic, and Guerrino's quest was to have a happy end. When at last he reaches Rome, the pope orders him to go and join the fight against the Turks, and when in due course Durazzo is retaken and the prison gates thrown open, who should walk out, looking exactly as the Hibernians had led him to expect, but his father and his mother!

'Going there' had most certainly paid off and no reader of *Guerrino* could have been in any doubt that what happened in Hibernia was truly 'amazing to say' and that St Patrick's Purgatory was definitely worth a visit. More sceptical readers, though, may possibly have asked themselves whether Ireland's 'fabulous' well really did serve the purpose it was meant to do. There was no denying, if Andrea was to be trusted, that it was successful as a missing persons' bureau or, judging by what Fazio said, as a mollifier of authorial *Angst*, or else, as in the case of Malatesta Ungaro, as a stamp of outstanding heroism. But what about its purging 'grace'? It is true that thanks to it Guerrino was returned to the fold to which he rightfully belonged and that, when he died, he was fully 'confessed and communicate'.[36] Yet it is also true, judging by most of the evidence that has come down to us, that the motives of the pilgrims who visited St Patrick's Purgatory had little to do, on the whole, with the hope of fast-tracking their admission to paradise. For instance, as we have just seen, the prime concern of Antonio Mannini was 'to recover [his] honour and state, and that of [his] house'. Is it any wonder that Pope Benedict, in Andrea's work, should ask Guerrino to 'come back to me to tell me what you found there'? He no doubt had his suspicions; and as a matter of fact he was not alone. Even the Archbishop of Hibernia was suspicious.

From the time of St Patrick himself it had been the tradition at St Patrick's Purgatory, according to Andrea, that pilgrims record their experiences in a specially designated book: 'at the time of Saint Patrick many people went in and many returned, and many remained; and for those who returned Saint Patrick had everything written down which they said they had seen and heard, so as to know how to instruct whoever went in'. Accordingly, Guerrino too was asked 'what he had seen, and everything was put in writing', but when the Archbishop of Hibernia was informed, 'to Meschino it did not seem that he believed it'.[37] To make matters worse, when Guerrino then reported back to the pope in Rome, the pope listened to him in stony silence, without saying a single word on the matter. And if that was not bad enough, when Chiericati visited the place, a century or so after Guerrino with the help of Andrea 'had made known to the incredulous the Purgatory of Saint Patrick', he not only treated the whole pilgrimage with incredulity, as we know, but he actually cast doubt on the reliability of the book in which 'Guarino of Durazzo' was 'listed'. Worse still, at the end of the sixteenth century, Guerrino's Patrician purge itself fell victim to a purge, so that, from then until today, *Guerrino il Meschino* was only ever published, as far as we can tell, without the chapters dealing with 'the pains of the Purgatory of St Patrick'. Purgatory and St Patrick's Purgatory had become too controversial.

The controversy, as a matter of fact, had started almost as soon as purgatory was 'born', for not only was its existence contested by the Orthodox and heretics, but with the growing influence of preaching orders, for whom purgatory became a tool of blackmail, so to speak, with which they bent the faithful to their will, by underwriting its soundness as a surety for the returns to be gained from alms-giving, it very soon generated intense opposition within the Catholic fold as well. Thus, for instance, in the second half of the fifteenth century, Masuccio Salernitano (1410?–1475?), a latter-day Boccaccio from the Kingdom of Naples, who shared the Neapolitan monarchy's suspicion of what it saw as the fearful and indomitable lust for power of the Church, dedicated much of his popular and violently anti-clerical collection of short stories, the *Novellino*, to lampooning the crimes of the clergy, suggesting at one point that the only way of putting an end to them and to the Church's hubris would be for God to 'soon destroy purgatory'.

That was also to be the view of Protestants, whose protests at the abuses committed by the Catholic Church were largely generated by the Church's shameless merchandizing of indulgences, and who would soon make it an article of faith that purgatory was a 'Romish' abomination. 'The Romish doctrine concerning Purgatory – says Article 22 of the *Book of Common Prayer* of the Church of England – is a fond thing vainly invented, and grounded upon no warranty of Scripture; but rather repugnant to the word of God'. In response to the criticisms of the Protestants, the Council of Trent (1545–63), which was to set in train what is known today as the Counter-Reformation, while reaffirming the validity of the doctrine of purgatory, nevertheless stipulated that it should be preached only in its 'healthy' essentials, as taught by the Church Fathers and the Councils of the Church. In particular it warned bishops, as guardians of the faith, that they should 'prohibit as scandalous and offensive to the faithful the preaching to the uneducated populace of anything that relates to simple curiosity and superstition, and that has the scent of filthy lucre'. It is possible that among those simple curiosities and superstitions the Church also intended to include St Patrick's Purgatory, from which, in fact, it had already sought to distance itself more than a century earlier. Or so at least it would seem.[38]

The story goes that in 1494 a canon from the Netherlands, 'driven it is not known by what spirit', sought and obtained the permission to wander the world as a 'poor mendicant' and that, having 'entered into various Christian lands and regions', in due course 'he came to the Kingdom of Hibernia, in order to see and even to enter into St Patrick's Purgatory, about which many things are told'. However, when he got there, not only did he find that 'entering into' was well-nigh an impossible feat, what with greedy officials and clerics trying to extort money from him at every stage of the process, even though he was penniless and doing so was tantamount to committing 'the plague of simony', but also that there was precious little to see, 'for he did not know that the ancient miracle, since the faith had been established, had ceased, but that the inhabitants of the place, for gain and money, still told those who came to the place that sins were purged there'. Scandalized, when he realized he had been tricked, he resolved at once to 'abolish this illusion of the simple-minded' and without delay made his way to Rome, where he reported everything to the pope's penitentiary, 'an honest enough and church-like man', who in turn reported it to the supreme pontiff, whereupon the latter, 'taking it badly that the simple-minded were being deceived in this way', immediately gave orders that 'that place, where once there was the entrance to the purgatory that is said to be of St Patrick be undone from top to bottom'.

The story is reported in an appendix to the life of St Patrick contained in the authoritative *Acts of the Saints* [*Acta sanctorum*], an enormous register of the lives and deeds of the saints officially recognized by the Catholic Church, which was compiled by Jesuit scholars and first published in 1735. The *Acts* give no source for the story, except to say that it is mentioned in 'some old manuscript', and despite various searches carried out by academics over the years, no documents have yet been uncovered in the Vatican or other archives which could lend support to it. The closest there is to a proof is a reference in the *Annals of Ulster* which states that

> the Cave of the Purgatory of St Patrick on Loch Gearg was broken this year by the Guardian of Donegal and by the representatives of the Bishop in the Deanery of Lough Erne by authorization of the Pope about the feast of St Patrick [...]: it being understood by everyone in general from the History of the Knight and other old books that this was not the Purgatory Patrick got from God although they were, everyone, visiting it.[39]

Unfortunately, the year in question is 1497 rather than 1494. However, there are indications that in the Franciscan library of the Sicilian town of Noto, a gem of (crumbling) baroque architecture not far from Syracuse, there used to be a work 'about the site of Hibernia and Thule and the customs of their peoples', seemingly written in 1494 by a local Franciscan, Giovanni Antonio Provina, following a trip to Ireland, where he appears to have been sent as papal nuncio. Provina, who then dedicated his work to the Spanish Viceroy of Sicily, Ferdinando de Acugna (1453–94), could therefore well have been the person who oversaw the execution of the pope's orders. Unfortunately his work was lost following the terrible earthquake of 1693, which razed much of Noto to the ground, bringing down the Franciscan library too, and no other copy of it is known to have survived. Nevertheless, despite the lack of hard evidence, the story regarding the closure of St Patrick's Purgatory is now generally accepted to be true, and hardly an eyebrow has been raised on account of its many improbabilities.[40]

What makes it most improbable is the fact that the person to give the order, whether in 1494 or 1497, would have been none other than Alexander VI (1492–1503), the simonist Borgia pope who made of his family name a byeword for papal immorality and corruption and who himself took great delight in deceiving the 'simple-minded'. According to a contemporary of his, Alexander 'knew all the ways of money', and according to Machiavelli in the *Prince*, 'he never did anything else, never thought of anything else except deceiving men'. True, his instincts were those of a Church reformer – in 1495, for instance, he issued a bull for the foundation of the University of Aberdeen, with the specific aim of tending to the education of the clergy in that area – and in 1497 those instincts were allowed to gain the upper hand, albeit temporarily, as he underwent a crisis of conscience following the assassination of his favourite son, the Duke of Gandìa, possibly at the hands of his second son, Cesare, the hero of *The Prince*. Still, the story quite beggars belief; and we should be all the more wary to believe it as we have Chiericati's testimony to disprove it.[41]

Whereas Chiericati would have had good reason to refer to anything which cast doubt on the authenticity of the pilgrimage, since he too, like the canon from the Netherlands, seems to have considered it to be no more than an 'illusion of the simple-minded', in his letter to Isabella d'Este he not only describes the ritual of the pilgrimage in great detail, but he makes no mention whatsoever of St Patrick's Purgatory ever having been 'closed' or 'broken'. That is all the more surprising if we consider that he was in fact a papal nuncio who, even though he deplored the 'bother' caused by the 'great faith' of the native Irish, which made it difficult for him to 'go through the streets there, because everybody rushed to kiss [his] clothes, hearing that [he] was a papal nuncio', and even though he preferred the 'civility' of the 'illustrious Earl of Kildare' to the manners of the lords who ruled the rest of Ireland and were 'little more honourable than our peasants', nevertheless gave credit to the Irish for their exceptional devotion to the pope, who 'they say is their king'.[42] A papal nuncio of the sort would surely have reported anything which was contrary to the interests or the teaching of the Church. Indeed, if St Patrick's Purgatory was no longer meant to be in existence, he would presumably not have gone to Ireland in the first place.

We must assume, of course, that he did go. The 'spirit' in which he wrote the letter and the admission that he did not actually dare to look inside the purgatory might lead us to suspect otherwise, notwithstanding his detailed account of, say, Dublin's flourishing trade or of the way the Irish dress. 'There are countless ships there', he says of Dublin, 'which export salted fish, hides, cattle and hobby horses, and import wines and many kinds of goods', while of the Irish dress he writes:

> The men wear a cloth shirt, died in saffron, down to their ankles, with shoes without stockings, a grey cloak on top and a felt hat on their head, and they use nothing else; all are shaven, except for their chin. The women are very beautiful and very white, but filthy. They use the same saffroned shirts, with a red cap on their head after the fashion of Carmagnola.[43]

Since he was the nuncio to the court of Henry VIII in London, it is easy to imagine that he might have met a good few people there who had themselves been to Ireland and could therefore provide him with enough information to satisfy Isabella's 'exceptional mind'. Certainly, if St Patrick's Purgatory was out of bounds for a papal officer but he went to visit it all the same, his transgression caused no harm to his reputation or his career. That, as it so happens, was to occur a few years later and be the consequence of a 'great faith' of another sort.

A much respected diplomat, besides England, Chiericati also paid visits to Spain, the Swiss cantons, Germany, France, the Low Countries, Portugal, as we have seen, and even to northern Africa. Sometimes his missions would be mundane, but often they concerned matters of the highest importance. In England, for instance, he was charged with raising support for the pope's proposed campaign against the French, as well as with sorting out disagreements between Cardinal Wolsey and the papal tax collector, Cardinal Castellesi. In Portugal he was to enlist the support of the king for a crusade to convert the Coptic Ethiopians to Roman Catholicism. But no mission was to be quite as important, or quite as disastrous, as the one he undertook in 1522 for pope Adrian VI (1522–23), the last non-Italian pope before John Paul II. Adrian sent him to the Diet of Nuremberg, where Catholics and Protestants were to sort out their differences, with the mandate of reforming the Catholic Church and eradicating

Lutheranism, no less. However, allowing his zealousness to get the better of him, Chiericati succeeded only in alienating the various parties present at the Diet, including the Catholic supporters of the pope, and having gone so far as to demand the head of four popular Lutheran preachers, he was ultimately left with no choice but to beat a hasty retreat, in greater fear of his life than he had ever been at St Patrick's Purgatory. Thereafter he never left Italy again, remaining primarily in Rome, and he was never promoted any higher than to the see of Teramo, a diocese in the Abruzzi mountains south of Rome. Yet despite his enforced retirement, he continued to enjoy the patronage of popes, first of Clement VII (1523–34) and then of Paul III (1534–49), and his diplomatic skills continued to be in demand. When in 1525 the envoy of the Duke of Muscovy came to Rome to discuss a possible alliance against the Turks, it was Chiericati who was asked to look after him, and two years later it was Chiericati once again who was charged with accompanying a second Russian delegation to meet the pope in Orvieto, where Clement had taken refuge after the sack of Rome by the troops of Charles V, and more importantly (as far as we are concerned) where the construction of the local St Patrick's Well, a sixty-metre deep water cistern, had just got underway. Yet whether Chiericati mustered enough courage to look into that one, or whether it evoked in him any memories of Ireland's fables, the story does not tell.[44]

In all likelihood, the 'spirit' of the time made it safer not to dwell on such a matter. It may have been all right, a few years back, for a papal nuncio to go in search of Ireland's fables, so long as they were seen to be 'of little account', or for the forebear of the Este to visit 'fabulous Hibernia', so long as he did not 'seem' to be too taken by its cave, and it may still have been all right to name a bottomless pit in a papal city after Ireland's patron saint, so long as the 'little old' man himself did not gain too much in stature; but it was not all right anymore to trifle with the otherworld. The times were changing, and purgatory, with its Irish proxy, was due for relegation. Tullia d'Aragona (c. 1510–56), a high class courtesan who wrote poetry, seeking to establish her credentials as a writer of repute, was still able to include St Patrick's Purgatory in her verse plagiarism of *Guerrino*, which was published in Venice in 1560; but she nevertheless felt the need to raise the tone and style of the story, and when it came out, it was advertised, on the title page, as a work to edify its readers, 'a work all filled with pious charity, with long exiles, with unsurpassable gallantries, with undaunted feats of war, with glorious virtue, with unblemished love, and with the utmost faith'.[45] The readers, though, were reluctant to be edified: the work was only published that once, and it was the original *Guerrino* which they continued to prefer. That too, however, was shortly to be given a nobler semblance, and by the turn of the century, as has been mentioned, Guerrino was to find his parents and made communicate again without the intercession of Hibernia's diminutive saint. The cave had now lost its grace and was too 'fabulous' by far. And in any case, had the stories ever told the truth about it?

Notes

1. 'Che di Scozia, e d'Irlanda, e d'Inghilterra,/E de l'isole intorno eran le schiere/Che quivi alzate avean tante bandiere' – 'Or guarda gl'Ibernesi appresso il piano:/Sono duo squadre; e il Conte di Childera/Mena la prima, e il Conte di Desmonda/Da fieri monti ha tratta la seconda' – 'la remota Islanda' – 'ogni terra [...] che là giace,/Nimica naturalmente di pace' – 'De le spelonche usciti, e de le selve;/Hanno piloso il viso, il petto, il fianco,/E dossi, e braccia, e gambe, come belve' (ARIOSTO (1959), pp. 89–91: 10.75, 87–9).
2. For more on this see Haywood (2003a), Haywood (2004). The literature on Ariosto is vast. As a useful, though somewhat dated introduction see Brand (1974).
3. 'Schivando a suo poter d'alloggiar male' – 'poi che di banda in banda/Vide gl'Inglesi, andò verso l'Irlanda;//E vide Ibernia fabulosa, dove/Il santo vecchiarel fece la cava,/In che tanta mercé par che si truove/Che l'uom vi purga ogni sua colpa prava' (ARIOSTO (1959), pp. 89–91: 10.73, 91–2).
4. On the pilgrimage to St Patrick's Purgatory, including the *Treatise*, see Haren and Pontfarcy (1988). On purgatory, and its link with St Patrick's Purgatory, see Le Goff (1984: 'literary birth certificate', p. 181; 'great intellectual and moral revolution', p. 214), entries on 'Purgatoire' in *Dictionnaire de théologie catholique* (1936: cols 1163–1361), and on 'Purgatory' and 'Purgatory, St Patrick's' in *New Catholic Encyclopaedia* (1967: pp. 1038–9). Not everyone agrees with Le Goff's theory, but it is still the most compelling account of the emergence of the belief in purgatory. As a sample of a different take on the matter see McGuire (1989).

5. '... et ecce terra inter circulum se aperuit et puteus maximus et profundissimus ibi apparuit revelatumque est beato Patricio, quod ibi quidam purgatorii locus esset, in quem quisquis vellet descendere, alia sibi poenitentia non restaret nec aliud pro peccatis purgatorium sentiret, plerique autem inde non redirent et qui redirent, eos a mane usque in sequens mane ibidem moram facere oporteret. Multi igitur ingrediebantur, qui de caetero non revertebantur' – 'Jesu Christe fili Dei vivi miserere mihi peccatori' (VARAGINE (1969), pp. 213–14).
6. 'Locus ille, quem conspicis, est infernus, in quo dominus noster Beelzebub habitat' – 'dixeruntque ei, quod illa civitas paradisus esset' – 'Tunc Nicolaus inde ascendens in loco, per quem ascendit [*recte?* descendit], se restitutum invenit et omnia omnibus, quae sibi contigerant, narrans post triginta dies in domino feliciter requievit' (VARAGINE (1969), pp. 215–16).
7. MANNINI (1918), p. 55. 'Io mi facea e feci più volte presto per venire: alcuna volta non c'era nave, alcuna volta non mi lasciava il vento, alcuna volta mi ritrovai sanza danari [...] come a Dio piacque ci capitò un Nobile Cavaliere d'Ungheria, il quale si chiama M. Lorenzo Rattoldi per andare al Purgatorio di San Patrizio. Visitando io il detto Cavaliere e parlato con lui per gran tempo, e veggendo il suo santo proposito e divozione, chè lui venia di capo della Christianità del mondo nella fine del mondo per salute dell'anima sua, e io veggendomi essere qui presso a detto Purgatorio et in tante fatiche e pene, considerando quale per adrieto era stata la vita mia, peccatore et sempre atteso alle cose mondane, confortato da buona speranza, mi volsi a Dio, da cui per sua misericordia spirato, diliberai e propuosi per salute dell'anima mia andare al detto Purgatorio, immaginando e prendendo speranza nella divina misericordia che forse per l'avvenire arei migliore ventura e grazia, e fine delle mie lunghe tribulazioni' (MANNINI (1886), p. 155).
8. MANNINI (1918), pp. 55–6, 59–61. 'Io nel Purgatorio serato rimasemi' – 'là dov'è il Purgatorio è un lago intra altissimi monti' – 'tutto quel lago è il più freddo paese dove fussi mai; su le montagne di Brigha non è freddo di verno rispetto di questo luogo' – 'molte tentazioni e visioni' – 'pregai Messer Iddio di 4 grazie, come il Priore m'insegnò dovessi domandare; che le tre per lo meglio mi taccio, la quarta domandai che piacesse alla misericordia di Dio che io potessi ricoverare mio honore e mio stato e di casa nostra quand che sia' – 'i capegli mi si levarono in capo e istavano arricciati ritti, che io non gli poteva colle mani quasi abbassare, nè fare stare giuso' – ''l cuore mi cominciò a ritremare et a battere' – 'sì lungamente come io pote' durare' (MANNINI (1886), pp. 156–7, 159–61).
9. 'Sì che questo è quanto ho trovato de le fabule, che se dice de l'isola de Hibernia et del Pozzo de s. Patrizio; le qual anche che sieno di pocho momento non ho voluto pretermetter de scriverle a V. Exc.ª come le promisi, conoscendo il suo peregrino ingegno, che non sol vol intendere le cose grande, ma anche le minime.

Sì che V.S. Illma accepti il tucto in quella parte, chel scrivo' (CHIERICATI (1873), p. 92). See also CHIERICATI (1987). On Chiericati see Foa (1980), Morsolin (1873).

10. 'Che in Italia valaria molto et saria in gran existimatione' – 'Et li villani ne attrovano tante, che è cosa mirabile, de le quale ne fanno bonissimo mercato. Et queste sono quelle perle, che si domandano scozese' – 'Lo aere è temperatissimo et più caldo, che non è in Anglia, che è assai mirabile' – 'grande religione' – 'Dicono, che il Papa è il suo Re' – 'diversi signori, che sono poco più honorevoli de li nri contadini' – 'valeno molto ne le arme, perchè sempre se esercitano in guerra fra loro' – 'Dicono, che noi siamo bestiali a far proprietà de beni de la fortuna et che loro vivono naturalmente, che ogni cosa deve andare in comune' (CHIERICATI (1873), pp. 88, 90–1).

11. 'Una abatia de canonici regulari' – 'una ciptà de Hibernia chiamata Dublino' – 'Ivi dal Rmo Archiepiscopo et da l'Ill. Conte de Childaria, vicedomino di quel isola, fossemo honorevolmente tractati' – 'et è homo da assai et ha tucta la civiltà de Inghilterra' – 'gente bestiale' – 'per lochi assai ameni sopra il mare' – 'deserta' – 'pieno de ladri et de boschi, lagi et paludi' – 'in un fazo incavato [...] pagasi uno dinaro per uno' – 'nove iorni e nove notti' – 'la maior penitentia' (CHIERICATI (1873), pp. 87–9, 91).

12 'Il qual loco non ho possuto veder per non haver voluto mirar drento, spaventato da le cose, che si dice; ma stetti lontano da la porta tre passi. Et li canonici introrno dentro con due facelle de pino, et io rimirai a diritura il tucto, che non si vede altro, che saxo' (CHIERICATI (1873), p, 89).

13. 'De quelli, che sono intrati me presente, dui hanno visto tali visioni, uno de li quali è uscito come insensato, et diverso modo interrogato ha ditto esser stato battuto gravissimamente et chel non sapea da chi: l'altro ha dicto esserli apparso diverse donne de bellissima forma, le quali lo invitavano a vrler [*recte?* voler] manzar seco et far bona ciera, apponendoli frutti, et cibi de molte sorte; et dice, che pocho li manchò, che non si lassasse vincer per la gran debilità, in la qual si trovava. Li altri dissero non haver visto cosa alcuna, nè sentito, se non freddo fame et gran debolezza' (CHIERICATI (1873), p. 90).

14. On Malatesta Ungaro's pilgrimage see Cardini (1990), Haren and Pontfarcy (1988). 'Vedi [...]/la coppia d'Arimino, che 'nseme/vanno facendo dolorosi pianti' ['See the couple from Rimini, who go together, producing woeful cries']: *Triumphus Cupidinis* III, 82–4, in PETRARCA (1996), p. 150. 'Coppia d'Ariminio' – 'Viola Novella, moglie de uno chiamato Coça Battaglia' – 'Malatesta signore di Rimino' – 'per lo suo amore andò al Purgatorio de sancto Patritio' (quoted in Alessio (1990), p. 180). On Ilicino see Haywood (1997), Merry (1986).

15. 'Sed nostra etas tertium addit nobilem militem Malatestam scilicet de Arimino, Ungarum cognominatum' (Silvestri (1955), p. 126); 'solum enim redisse scribitur patricius et Nichola, quamquam nostro seculo Strenuus miles Malatesta de Arimeno cognomento ungarus ad iactantiam vulgo dixerit a dicto puteo se redisse' (Bandini (MS), f. 45v).
16. '... Servi pauperis non contemnere leve munus [...]. Rarus quippe est hac nostra tempestate qui [...] non triste otium proferat actibus gloriosis [...]. Te amor iste precipuus ad extremum vehexit occasum, ubi saltuosa montium avia quamque invictus laboribus peragrasti et rapacissimi amnes ripis undantibus superati sunt. Et quod grave nimis esset consuevit nobili delitiis enutrito, non inedia, non vigiliis, non algore, non siti, hec omnia sepissime passus [fuisti] succubuisti quamquam pedes Hybernie deserta calcaveris, raris admodum inhabitata pastoribus. Ubi facies orrenda loci illius et squallida regio omni prorsus amenitate nudata, terribili frequentatione ferarum humanos terret aspectus. Quid dulces portus? Quid vada neptunia? Quid hyemes referam et iratos fluctus Occeani? Quotiens non absque gravi capitis discrimine tibi sulcata sunt ut ceptum iter expleres. Postremo quia longum esset per cuncta discurrere, ista pretereo, ut Anglia, nunc rege felix et dextro sydere propagata, ut semoti fines Britanie te viderunt et ultima Tile, necnon cum vespere celsus Athlas et pene totius orbis occidui regiones. Quid? Postea locus ille cavernosus et opacissimus antris abditus recepit te volentem. Ubi nox densa caligine et nubes tenebrosa circumdat nigros hyatus Inferni. Illinc tamen intrepidus rediisti unde rarus intrantum numerus remeavit, tantus error ut autumo mortales alligat per cecos meatus et tristia latebrarum. Teret auditus humanos delitentis putei inexpleta vorago [...]. Vale in longum regalis militie, clarum decus, me tuis ascribendo clientulis' (Checco (1984), pp. 16–19). 'Piacciavi, generosa Erculea prole [...] aggradir questo che vuole [...] darvi [...] l'umil servo vostro' ['May it please you, noble son of Hercules [...] to accept this, which your humble servant wants to give you']: Ariosto (1959), p. 1: 1.3.
17. '... lasciò andare il mestier del corpo [...] acciò che quando serà quel tempo, io truovi e 'l segno e non affoghi nella calca [...] una orazione alla nostra Donna che gli facesse grazia, raccontando tutti i luoghi che oltre mare avea vicitato' (Sacchetti (1996)). Dolcibene's oration is in Tortoli (1904), pp. 5–15. 'Heret calamus in manu trepida, cor tepescit et nutat animus dum tuas tentat laudes aggredi, Heros magnanime' (Checco (1984), p. 16).
18. 'Da una banda del scopulo è forato, et per un piccol buco si pone et ritoglie un vaso per le necessità loro. Et sempre sta a quel buco uno de li canonici che lì predicano, che siano costanti, nè si lascino superar da le diaboliche tentationi, perchè come dicono li appareno de varie et innumerabili visioni, et assai

de loro usciscono smemorati et pazzi, qual dicono haver cesso a le tentationi' (CHIERICATI (1873), p. 89).

19. 'Il iorno seguente uscitero mezo morti et furono restorati melio, che si potè, et furono posti sul libro, che sta in Chiesa, nel qual sono descripti tucti chi vano. Il primo descripto fu Guarino da Durazzo, qual io cresi esser fabule. Ma adesso ivi l'ho visto descripto antiquamente in un libro pergameno' (CHIERICATI (1873), p. 90). For more on Checco see Picard (1988).

20. On Andrea da Barberino see Allaire (1993, 1996, 1997, 2006), Cardini (1983: *Guerrino* read in the USA, p. 186), Osella (1932). There is a modern critical edition of *Guerrino* (BARBERINO (2005)), but I decided not to quote from that, since, by the editor's own admission, it does not give the true flavour of the text as it was read in the fifteenth-sixteenth centuries.

21. 'Secretamente vna notte stette ingenochiato a lato al santo Sepulchro de Misser Iesu Christo & pregollo che gli desse gratia di trouare el suo padre e la sua generatione: & giuro sopra la santa sepoltura de attendere la promissione ad Antinisca & che la farebbe christiana se trouasse el suo padre' (BARBERINO (1534), p. 53v).

22. 'O maledetto seminatore de scandolo: la diuina iustitia dia a te degno merito delle anime che hai fatto e fai perdere per la tua operatione' (BARBERINO (1534), p. 45r).

23. 'E per penitentia li impose come lui hauea hauuto ardire contra el comandamento de la lege de dio de intrare doue staua la Sibilla & de andare a visitare li idoli cioe li arbori del Sole: cosi volea che per comandamento lui andasse a lo purgatorio de sancto Patricio el quale e sotto larciuescovo de Iberina (*sic*) in lisola ditta Irlanda: e prima al barone misser sancto Iacobo de Galicia e a Santa Maria de finisterre: e lui molto lo ringratio quando li disse fatta questa penitentia tu sei absoluto da Dio e da mi. E pregollo el sancto padre dicendo se tu entri in lo purgatorio de sancto Patricio se tu poi torni a me a dire che cosa tu hauerui trouato: ma questo non te lo comando anci lo remetto ala tua uolunta: e lui rispuose o sancto padre ogni cosa faro se tanto uiuero chio ci gionga: ma sol vna cosa me impaza che me mancha dinari per spendere per lo camino: e lui li fece dare ducento dinari doro [...] e questo fu papa benedecto terzo nelli anni del nostro Signore Iesu Christo octo cento vintiquatro: & era imperatore Carlo magno el Vechio' (BARBERINO (1534), p. 91r).

24. 'Io vegno de tutto lo mondo e non so donde vegna ne doue me vada [...] cercasti mai el mondo: respose lhosto sono stato in Soria in Romania in ponente in spagna in ingilterra & in Flandra e hora son tornato in la patria mia ho prouato del bene e del male e se hauero mai figlioli grandi che se possano guadagnare le spese io li faro cercare del mondo perche chi non ha cercato del mondo non e homo' (BARBERINO (1534), p. 80v).

25. 'Guerino prenominato Meschino. In questo libro se tratta una hystoria breue de re Carlo imperatore. Poi del nascimento di quello famosissimo caualliero nominato Guerino. Et delle grande battaglie che lui fece con Turchi e Saracini & con molto altre generatione di gente: Et della diuersita de animali che lui trouo andando alli arbori del sole: & de l'habitatione della Sybilla e suoi inganni: anchora narra delle pene del Purgatorio di santo Patricio. Per loquale se po intendere la situatione delle prouincie e Citta: e della diuersita delli huomini e li loro costumi' (Barberino (1534)).
26. 'Fecel imparar lenguagi per l'utilita de la mercantia e per nauegare: imparo turchesco' – 'E gionto in ibernia domando de larciuescouo e non sarebbe stato inteso se non hauesse saputo bon latino e bona grammatica' (Barberino (1534), pp. 11v, 93r).
27. 'E tutto el suo viazo haueua fatto scriuere dal zorno che lui fu fatto schiauo in Costantinopoli fin ala partita de Ingheltera' – 'el.M. subito scrisse a constantinopoli ad Alexandro e scrisse in persia, a la citta de presopoli ala bella Antinisca facendoli secretamente a sapere come lera viuo & hauea ritrouato el suo padre & scrise in babilonia: e in barbaria per fare fede che lera stato per tutte le parte che diceua. Anchora scrisse in la morea & in ingheltera a misser Dinoino & in poco tempo furono verificate le sua parole per vero hauere quasi tutto el mondo cercato per trouare la sua generatione' (Barberino (1534), pp. 106v, 112r).
28. 'Molte castelle e ville e molto temperato paese e molto viueno le persone: ma perche le lontano paese dala terra ferma e male habitato. E questo demonstra la humana natura che per consuetudine participano piu de terra che altro elemento: e con tutto che in questo paese se viue molto: nondimeno habitano piu in terra ferma perche la terra monstra essere piu naturale elemento & vna nostra madre' – 'verso la Spagna' – 'molti bestiami' (Barberino (1534), p. 93r).
29. 'Verso ingheltерra' – 'vltime citta del mondo' (Barberino (1534), p. 93r). 'Molto è il paese alpestro e pellegrino/e la gente v'è ruvida e salvatica,/aspera e fiera a ogni suo vicino:/vero è ch'egli han mutato vezzo e pratica/per bontá d'Adoardo, ch'è or vivo,/che gli ha frustati piú su che la natica' (Uberti (1952), vol. 1, p. 328: IV, xxvi, 19–24); 'rege felix' (Checco (1984), p. 17); 'ha tucta la civiltà de Inghilterra' (Chiericati (1873), p. 92).
30. 'Tutta la humana natura e bellissima. citta grande & bene populate sie bene cinquanta mia di bellissimo paese bene habitado e de ogni cosa diuitioso e si e fortissimo paese si de terreno come de gente e belle forteze e sempre stano su le porte a bona guardia per li corsari. E signoregia questo paese larciuescouo de ibernia & hanno cosi mogliere li sacerdoti come li seculari. & e beati quelli che se possono aparentare con loro' (Barberino (1534), p. 93r).

31. 'Laquale isola alhora era chiamata lisola sancta perche non ce morite mai persona tanto e laere temperato & non glie suso nissuna cosa venenosa ne lupi ne cosa pessima: e questa isola se chiama secondo li antiqui isola suuentioni: non li po viuere suso ne volpi ne fuini ne molti altri animali che sono sopra la terra: & glie laere cosi ben temperato e ben disposto che li homini e le donne li vengono in tanta vechieza che la vita li rincresce a loro: e quando se vanno cosi inuechiandosi se confessano e conzasi de lanima e fannosi portare in mare oltra a questo streto di mare che e tra Irlanda e questa isola come sono in irlanda subito moreno. E questo ordine sie conceduto a loro da sacerdoti: questa isola per antiquo fu chiamata isola sancta: e de presente lisola carnara. Elquale nome li hanno posto li marinari quasi volendo dire che son carnifici de loro medesimi perche portano li loro vechi alla morte. Questa gente viuono santamente e sono amici de dio e molto religiosi. non li sono tanti ladroni ne tanti pessimi inimici de Dio e di sancti quanti sono tra noi: & se cosi li marinari hanno posto loro questo nome: de dime ti che cosa e hoggi al mondo che christiani non lhabino guasta. & e pieno el mondo de ogni ladronezo e de ogni tradimento e de ogni fraudulentia. quale padre: quale fratello: quale religioso se po hogi fidare luno con laltro. che religione: che amore: che fede: che charita? nissuna: Tutti li homini hanno in odio li boni & agradeno li catiui. La citta di fiorenza non e gia quella isola sancta. ma per opposito ha ben consuetudine el morire ma non la prouidentia del morire: e di questo mi rincresce che nissuno non ce po viuere in questa isola' (Barberino (1534), pp. 93v–94r).

32. 'The island is divided into two parts, the largest being situated between west and north and called in the Irish tongue "Kernagh", which in the Latin language could be said to mean the Isle of Clamour' ['Dividitur [insula] enim in duas partes, quarum maior inter occidentem et septemtrionem optinet locum suum Kernagh hiberniali ydiomate nuncupata latino quidem sermone Clamoris Insula dici potest'] (Delehaye (1908), p. 47). On the *Visions of George* see Haren and Pontfarcy (1988), pp. 120–68.

33. 'Quelli che al mondo se haueano fatto adorare' – 'tu manifesti ali increduli el purgatorio de sancto patricio: & le pene de lo inferno & li altri ministeri [*sic*] che tu hai veduto e vederai' (Barberino (1534), p. 105r). Dante's typology of sins in hell, in increasing order of gravity, is: lust, gluttony, avarice and prodigality, wrath, heresy, violence, simple fraud, treachery. In purgatory, in decreasing order of gravity, it is: pride, envy, wrath, sloth, avarice, gluttony, lust.

34. 'Li primi christiani non furono gente che se curasseno de viuere senon in vanita ma non che attendesseno a crescere la fede catholica: in questo non se dauano tropo fatica de fare batizare alcuna persona. Essendo sancto patricio religioso & amico de Dio' (Barberino (1534), p. 94r).

35. 'Da vna porta venne due persone vno homo & vna donna lhomo hauea grande capigliata biancha & auolupata la barba folta e grande: e tutti li panni straciati: la femina tutta pelosa: descalci li piedi: le onghie longe tutte le gambe rognose de piu parte monstraueno le carne e cosi era la donna come lhomo. Disseno a lui quelli sancti guarda bene costoro: impero che tu trouerai el tuo padre e la tua madre in questa medesima forma: e lui li guardo denanci e de dreto per cognoscerli e domando costoro come haueano nome e non resposeno' (BARBERINO (1534), p. 106r).
36. 'Confessato e communicato' (BARBERINO (1534), p. 139r).
37. 'Al tempo di sancto Patricio introrono molte persone e molti tornorono: e molti si romaseno: & de quelli che tornorono sancto Patricio faceua scriuere ogni cosa che loro diceuano hauere veduto e sentito per sapere amaestrare cui lintraua' – 'lo domandorono de quello che lui hauea veduto. & ogni cosa fu misso in scritto' – 'al Meschino non parue che lui credesse' (BARBERINO (1534), pp. 94v, 106v).
38. 'Che Idio possa presto destruere il purgatorio' (MASUCCIO SALERNITANO (1979), p. 20: novella II, esordio). *The Book of Common Prayer*, p. 304: Articles of Religion XXII (www.vulcanhammer.org/anglican/bcp-1662.pdf). 'Praecipit sancta synodus episcopis, ut sanam de purgatorio doctrinam, a sanctis Patribus et sacris conciliis traditam, a Christi fidelibus credi [...] Apud rudem vero plebem [...] ea vero quae ad curiositatem quandam aut superstitionem spectant, vel turpe lucrum sapiunt, tanquam scandala et fidelium offendicula prohibeant' (*Dictionnaire de théologie catholique* (1936), col. 1279).
39. 'Quo spiritu nescitur ductus [...] pauper mendicus [...] diversas mendicando Christianorum patrias et regiones ingressus est, venitque tandem in regnum Hiberniae, ut videret et etiam intraret Purgatorium S. Patricii [...] propter lepram simoniae [...] nesciebat enim quod antiquum miraculum, jam fide firmata, cessaverat; verumtamen incolae loci, ob quaestum et nummos, purgationem peccatorum inibi adhuc fieri advenientibus asserebant [...] illusionem hanc simplicium aboleri cupiens [...] viro satis honesto et ecclesiastico [...] graviter tulit taliter simplices decipi [...] ut locum illum, in quo quondam introiutus fuerat ad Purgatorium, quod S. Patricii dicitur, funditus everterent' – 'ex quodam vetusto codice' (*Acta Sanctorum* (1735), p. 590). On this see Haren & Pontfarcy (1988), pp. 190–201, Haywood (2002).
40. On Provina see Littara (1997), pp. 84–6. 'Alexander VI's order [...] to close the cave was carried out to the letter', says the *New Catholic Encyclopaedia* (1967), p. 1039.
41. 'Pecuniae omnes vias novit' (Sigismondo de' Conti (1883), vol. 2, p. 270); 'non fece mai altro, non pensò mai ad altro che ad ingannare uomini' (*The Prince*, chap. 18: MACHIAVELLI (1976), p. 99). On Alexander VI and the Borgias see

Cloulas (1987), Mallett (1969). On the foundation of the university of Aberdeen see De Roo (1924), vol. 4, p. 456.

42. 'In quel loco non potea andare per la strada, che ogni uno correa a basarmi la veste, intendendo che era nuntio del papa, adeo che quasi bisognava star in casa per forza; tanta era la molestia sua, che nasceva da una grande religione' – 'Dicono, che il Papa è il suo Re' (CHIERICATI (1873), pp. 90–1).

43. 'Vi sono de infiniti navilii, che asportano de li pesci salati, corii, bestiami et cavalli ubini, et portano ivi vini et merce de molte sorte' – 'Il vestir de li homini è una camisa de tela, insofranata fino nel collo del piè, con scarpe senza calce, una Bergna sopra de griso et un capello in testa de feltre, non usando altro; vano rasi tutti, salvo chel mento. Le donne sono bellissime et candidissime, ma sorride [*recte?* sordide]. Usano le medesime camise insofranate con un berretton rosso in capo a la carmagnola' (CHIERICATI (1873), pp. 87, 91).

44. Orvieto's 'pozzo di san Patrizio' is the work of Antonio da Sangallo il Giovane (1484–1546). For more about it see Palazzetti (1998).

45. 'OPERA TVTTA PIENA DI PIETOSA CARITA DI LVNGHI esilij, di cortesie insuperabili, d'inuitte & bellicose proue, di gloriose uirtù, d'Amore inuiolabile, & di somma fede' (ARAGONA (1560?)). For information on Tullia d'Aragona see Allaire (1995), Kristeller (1980), Pallitto's introduction in ARAGONA (2006).

CHAPTER FIVE

Divided World
Imagining Ireland(s) at the Time of the Reformation

Authors and works discussed in this chapter (in chronological order):

- GIOVANNI PONTANO (1422–1503), *Actius* – also see Conclusion
- POLIDORO VIRGILI, aka POLYDORE VERGIL (c.1470–1555), *The History of England* [*Anglica historia*]
- LUDOVICO ARIOSTO (1474–1533), *Orlando Furioso* – also see Chapter Four
- PAOLO GIOVIO (1486–1552), *Description of Ireland* [*Descriptio Hyberniae*]

If stories do not tell the truth, neither, probably, does history, and even less so in Italian, since Italian, as we said in the Introduction, does not distinguish between the two. Both are *storia*. More importantly, however, both are language-games and, like all games but especially, as this chapter will confirm, those favoured by humanists, language-games are played to be won, and winners take all. The truth is *theirs* for the taking. Unless, that is, there is someone who knows better, someone who can deliver the Truth from on high.

'If you want the truth not to be hidden from you, turn the whole (hi)story upside down: that the Greeks were routed, and that Troy the victor was, and Penelope a whore': so says St John the Evangelist, speaking on the moon to Prince Astolfo, who has gone thither to retrieve the lost wits of his lunatic cousin Orlando and who, as a bonus, receives advice from his saintly guide about the true meaning of life. I know what I am talking about, St John hastens to add, 'for in your world I too was a writer'.[1] The truth, in other words, is that all writers lie. But it is in fact a deceptive kind of truth. If it is indeed the case that all writers lie, but if it is a writer who tells us so, does that mean that, as he tells us so, he is actually lying and therefore that, as he lies, he is in fact telling us the truth?

The conundrum lies at the heart of Ariosto's *Orlando Furioso*, where all of this takes place, and it is Ariosto's way of challenging his readers to reflect on the seriousness, or otherwise, of literature, and on the vexed question of what we today would call freedom of expression. Who has the right to decide what a story means? – is what Ariosto asks us to ask ourselves. Who can control the meaning of stories? Who are the most trustworthy guardians of the narratives within which our lives unfold and make sense? Who owns language, and who is free to subvert it? Or put simply – since language equals power – who are the rightful masters of the word and of the world?

These questions are put to us repeatedly in the course of the poem, or rather they are put repeatedly to Ariosto's patrons, in an incessant game of cat-and-mouse which aims to bolster their authority while at the same time undermining it. 'Not so pious was Aeneas', St John explains to Astolfo,

> nor so strong Achilles
> as fame would have it, and Hector not so proud;
> thousands upon thousands upon thousands have there been
> who could in truth outmatch them,
> but it was the palaces and noble mansions gifted
> by their heirs which granted them
> such lofty honours without end,
> thanks to the honoured hands of writers.[2]

Put otherwise: without *your* generosity, my lords, there might be no stories, but it is *we* writers who create the stories. It is up to *us* to make and unmake reputations, to construct and deconstruct history. Or is it?

The most memorable episode in this (deadly serious) language-game occurs just after Ruggiero has left 'fabulous Hibernia' and it involves a peerless Irish king who was especially created for the third, revised and enlarged edition of the poem, which came out in 1532 and was thereafter to be the standard version of the work. Having seen Hibernia, Ruggiero flies towards the island of Ebuda (an island north-east of Ireland, which figures in Ptolemy's *Geography*) and peering down from his flying horse he sees a damsel in distress tied naked to a rock and about to be devoured by an orc. She is Princess Angelica, the most beautiful woman in the world and the object of every knight's desire, and she is a victim of the cruel custom of the Ebudans, which requires them to provide the monster which infests their seas with a continuous supply of young and living human flesh. Moved by her distress, as every knight should be, Ruggiero resolves to rescue her; but for want of staying power, instead of killing the monster he simply stuns it, with the help of the magic shield he is bearing, and for lack of principle, instead of doing what a knight should do, when flying off with the rescued damsel seated (naked) on the rump of his horse, he decides to carry her to a secluded wood and try his chance at 'mounting' her. For her good luck Angelica has a magic ring which renders you invisible if you hold it in your mouth, and Ruggiero is clad in his armour, which it is impossible to slip off as you would a pair of jeans. For Angelica, therefore, all is well that ends well. But not for Ruggiero or for Ariosto's patrons. Ruggiero, 'worthy of eternal glory', is meant to be their heroic ancestor. His destiny is not to prey on defenceless damsels but to marry his betrothed Bradamante and with her to give rise to the noble race of Este who shall 'return Italy to her erstwhile honour'.[3] It is clear, however, that for history to take that course both he and they will need to be taught a thing or two about chivalric *savoir-faire*, and about the 'honour' which the 'hands of writers' can bestow and may expect. It is that which Hibernia's peerless king was conjured up to do.

In the 1532 edition of *Orlando Furioso*, within just a couple of cantos of Angelica's escape from Ruggiero's lustful clutches, Ariosto decided to include a second take on the damsel-tied-naked-to-a-rock episode, which is itself a rewriting of the ancient myth of Andromeda and Perseus. In this case the damsel is Olimpia, a princess from the Netherlands, and her rescuer is Orlando, who manages to slay the monster single-handedly, by rowing into its cavernous mouth, propping it open with the anchor from his boat, and chopping and charging with his sword until the monster has spilled all its blood and given up the ghost. Orlando, though, is not alone, for even as he does battle with the monster and then takes on the Ebudans themselves who, far from being grateful, want swift revenge for his having dared to meddle with their 'vain religion',

without resistance those of Ireland,
from several sides came to the island:
and their pity spent, a brutal slaughter made
of that people on all shores;
be it justice or be it cruelty,
neither sex did they consider nor age.
[...]
The belongings were sacked, fire set
to the homes; the plebs was killed.
The walls were all razed to the ground,
not a single head was left alive.[4]

The Irish have come to deal once and for all with the problem of their bothersome neighbours, the Ebudans; but as we can see, they do so very much in the manner of the 'hairy beasts' we had imagined them to be following the review of the troops outside London. They have no pity, they slaughter brutally, and they cannot tell the difference between cruelty and justice.

Amazingly, however, the leader of these 'beasts', Ireland's King Oberto, is a paragon of chivalric virtue, who does not put a foot wrong when dealing with a damsel in distress. Having reached the shore where Olimpia stands saved and naked, he feasts his eyes on her, falls in love with her, gets new clothes for her, takes her home, defeats her enemies, and finally marries her and turns her from a countess into a 'great queen'. Not only is Oberto a perfect knight, though, he is also a close friend of Orlando's, whom he recognized at once, even though Orlando did not look himself, being drenched from head to toe in blood and water. The two had met at the court of Charlemagne in France, where Oberto had spent his youth as a 'page of honour', and they had only parted company a year before, when Oberto had returned to Ireland to take the crown, following his father's death. Such good friends are they that Orlando then insists on accompanying Oberto and Olimpia to Ireland, in order to be present at their wedding, and though 'hardly a day did he remain in Ireland' and 'no prayers served to make him stay some more', before hurrying on he found the time to give advice to his Irish *protégés* on how to make their marriage a happy-ever-after one.[5] Yet it is not entirely clear what his motives could have been for doing so. We cannot tell for sure whether he was more concerned for Olimpia's safety, or for his own destiny (this was the second time he had to put his mission on hold in order to save Olimpia), or possibly for the stability of Ireland. What cannot be in doubt, though, is the subtext of the entire episode, coupled with the first take on the Ebuda story. My lords of Este – Ariosto signals to his patrons – please take note! A cavalier treatment of damsels in distress, and their affairs of heart and state, no more a glorious hero makes than does the mistreatment of 'the honoured hands of writers'. You would do well to remember that we writers have the power to lie. We can rewrite (hi)stories at will, so as to align the truth of the past with the truth of the present, or vice versa. We can always find some peerless Irish king to outmatch a misbehaving Este.

It is now generally believed that the additions which were made to the third edition of *Orlando Furioso* were meant as a reply to critics who had accused Ariosto of having written a poem that was too frivolous for the dramatic times in which they lived.[6] The Olimpia episode is certainly dramatic. Before being fed to the sea-monster of Ebuda, Olimpia had sawed off the head of her first husband (the son of the king of Frisia, whom she had married against her will) and then been betrayed by her second one (her true love, Bireno, Duke of Zeeland). More dramatically still, she had had as her enemies the wicked Frisians who, having killed her father and her brothers, were intent on depriving her of her subjects and her state (the County of Holland) and who, by decree of poetic licence and of Beelzebub, were in possession of deadly firearms, five full centuries before they had been invented (in Ariosto's own day). On that occasion too she had been saved *in extremis* by Orlando, who with the sheer power of his sword and his courage had managed to defeat the Frisians on his own and thereupon had seized all their weapons and hurled them to the bottom of the sea, in the hope of consigning them for evermore to hell, from where they had emerged. Yet it was all to no avail – the narrator then adds – for in our own day the Enemy of all Mankind, with no less cruelty than he showed 'when Eve deceived us with the apple', caused them to be found again by a necromancer and in due course to infest every part of the world, prompting the narrator to exclaim, in a heartfelt *cri de coeur*:

> Oh wicked and ugly invention,
> how did you ever find place in a human heart?

Because of you military glory is destroyed,
because of you the profession of arms is without honour,
because of you valour and virtue are diminished,
and often the evil seem better than the good.[7]

The message could not be more explicit and it makes of the entire episode one of the most emphatic commentaries in the poem upon the drama of Ariosto's age, providing an uncompromising morality tale for all who would be truly 'glorious'. It teaches how to fight and how to rule, how to love and whom to love, how to deal with one's enemies and how to treat one's friends. In short, it teaches how to be a knight whose armour truly shines. And since the polish on the armour is partly made in Ireland, one cannot help but suspect that the moral of the tale is also about that. About removing 'fabulous Hibernia' from the realm of fables and of trifles, and inducting it into Truth and History. About no longer leaving it suspended between two worlds, but bringing it full square into this one.

In 1532, much more so than in 1516, when the first edition of *Orlando Furioso* had come out, this world was a very divided world. Not only were the two superpowers of the day, Spain under Charles V (1500–16–58) and France under Francis I (1494–1515–47), locked in bitter conflict for control of Europe, but in the German lands the angry protests had flared up which were soon to pit Catholics against Protestants, on top of which England's Henry VIII (1491–1509–47) was on course to deny the authority of both pope and emperor (that same Charles V who was also king of Spain) by repudiating his infertile wife, Catherine of Aragon (Charles V's aunt), and making of himself the supreme religious and imperial leader in his kingdom. Against this background, of which Ariosto and the Ferraresi were only too aware, it is tempting to view the invention of Oberto as Ariosto's way of indicating that he, for one, was aware that beyond the Germanic Low Countries and behind England, there was a friendly, albeit 'beastly' Catholic country which could be counted on to lend assistance to Empire and Church in their hour of need – by, say, reining in feckless rulers (as the ruler of Holland, Olimpia is dangerously and selfishly ineffective), so that they do not distract the paladins of the Christians from accomplishing their mission. Unfortunately, there is no real evidence to prove that that is indeed what Ariosto thought, and as far as we know it was only in the

1540s that Catholic powers began to view the Irish in that light. What we do know, however, is that it was at the very time Ariosto was working on such additions to the *Orlando Furioso* that Italian writers, at least one of whom was a friend of his, were devising new ways of describing, or imagining, the Irish, with the aim, precisely, of aligning the truth of Ireland and its past with the truth of Europe's present.

In 1534, within two years of the publication of the third edition of *Orlando Furioso*, the first edition was published in Basel of what was to become a best-selling history of England, titled *Anglica historia*. Begun in about 1506 and with its first draft complete in 1513, the *Historia* was the work of Polidoro Virgili (Polydore Vergil: c.1470–1555), a humanist priest from Urbino sent to England as assistant papal tax collector, and it had originally been commissioned by England's first Tudor king, Henry VII (1457–85–1509), with a view to 'justifying' the Tudors before 'the courts and intelligentsia of Europe'. It consisted of twenty-six chapters, covering events up to the death of Henry VII, and in the chapter on Henry II, with a postscript in the chapter on Henry VII, it contained a sizeable description of Ireland. In 1546, in Basel once again, the work was published in a revised edition, still containing twenty-six chapters but with the description of Ireland, amongst others, significantly rewritten, and in 1555 a third edition was brought out, also in Basel, with an extra chapter on the reign of Henry VIII (up to 1537) and with further changes to the description of Ireland and to other parts. Thereupon the *Historia* was republished a further seven times – three in Basel (1556, 1557, 1570), one in Ghent (1556), one in Douai (1603) and two in Lyon (1649, 1651) – and before long it was acknowledged as the most authoritative history of England, both in England and on the Continent, and remained so for centuries to come.[8]

In the process Virgili's description of Ireland also came to be regarded as authoritative, although it was soon to share that distinction with a description of Ireland composed by Paolo Giovio (1486–1552), another humanist priest from Italy (Como, in his case) and another of the day's foremost historians (in his case authoring a universal history of modern times, from the French invasion of Italy in 1494 to the Council of Trent). The first part of what was intended as a universal cosmography, Giovio's *Description of Ireland* [*Descriptio Hyberniae*], which is accompanied by a description of

Britain (i.e. England), Scotland and the Orkneys, was in the process of redaction at the very same time Ariosto and Virgili were retouching their own accounts of the place, but it did not see the light of day until 1548, in Venice. It was never to enjoy quite the same degree of popularity as either *Orlando Furioso* or the *Anglica historia*, but it was nevertheless republished a further three times (1561, 1571, 1578), all of them in Basel, and it too was to be read, as far as we can tell, by Europe's courts and intelligentsia. Like the *Anglica historia*, it too was written in Europe's contemporary *lingua franca* (i.e. Latin), and like Virgili's description of Ireland, it too bore the stamp of humanism's new learning, which would have made it particularly appealing to the day's reading élite.

Whether Giovio and Virgili ever met, is not really known, but what we do know is that, directly or indirectly, they were both acquainted with Francesco Chiericati. When Chiericati arrived in England, his first task had been to secure the release of Virgili from the Tower of London, where Virgili had been locked up for having dared cross swords with the Archbishop of York, Thomas Wolsey (c.1471–1530), over the latter's machinations to secure a cardinal's hat for himself. Some twenty years later, Chiericati took part in the triumphal procession held in Naples to celebrate the victory of Charles V in the crusade against the ruler of Tunis, the formidable corsair Barbarossa (d. 1546), and on that occasion Chiericati's task was to be the bearer of the keys to Barbarossa's treasury which formed part of Charles's war booty. In due course the keys would pass into the possession of Alfonso d'Avalos, Marquis del Vasto (1502–46), a diplomat and soldier who was one of Charles's chief agents in Italy, and subsequently d'Avalos gifted them to his close friend Paolo Giovio. The gift must undoubtedly have pleased Giovio, who was a keen collector of the memorabilia of important men, but it was another set of keys on which he would much rather have laid his hands. In 1540, reeling at the disappointment of having yet again been passed over for promotion, he complained to a friend that someone, surely, must 'have snatched from the girdle of the good pope the key of the chest of the red hats and thrown it in the well of St Patrick'. Unfortunately, St Patrick kept the key concealed for evermore and the highest Giovio ever climbed on the ladder of ecclesiastical preferment was, like Chiericati, to a bishopric in a minor diocese of southern Italy; but as we shall see, he did not bear a grudge towards St Patrick or the Irish for the disappointment. Admittedly he harboured even greater doubts than did his friend Ariosto about the faculties of Ireland's 'little old saint', and in all likelihood the well he invested with the power to thwart his ambition was not the one in Ireland but its Italian surrogate, the water cistern in Orvieto.[9] Still, acknowledging the nuisance factor of Ireland's patron saint was a compliment of sorts. At the very least it meant that Ireland figured in the current lexicon of Vatican insiders, and as we shall soon find out, Ireland's nuisance factor, albeit with an unexpected twist, was to form the kernel of Giovio's portrait of the Irish.

If anyone was versed in the lexicon of the Church's inner circles, that person was most certainly Paolo Giovio. Although in the course of his life he kept returning to his beloved birthplace of Como, and although at the end of his career, when it was clear that St Patrick would never give up that key again, he chose to settle in Florence, as a guest of the Medici rulers of the duchy, his favoured stamping ground were the Roman corridors of power, which he knew like no-one else. Having studied medicine in Pavia and in Padua, he moved to Rome at once and quickly made his way to the very heart of the Church administration, first serving as physician to a cardinal from Genoa and lecturing in the city's university, and later hitching his fortunes to the rising star of cardinal Giulio de' Medici (1478–1534), the cousin of Pope Leo X (1513–21); and when Giulio himself was elevated to the tiara in 1523, as Clement VII, Giovio became a permanent member of the pope's official household, and took up residence beneath the Sistine Chapel, in what was known as *Paradiso*. His appointment was as physician to the pope, but it was as an historian and conversationalist that his services were most required: his conversation, it would seem, was quite exceptional, and the art of rendering their name immortal was even more important to those who craved to leave a mark on history than the art of tending to their mortal body. When Clement passed away, in 1534, Giovio then used those gifts to gain attention from a galaxy of patrons, and in due course they brought him into the orbit of Alessandro Farnese (1520–89), the beloved grandson of Pope Paul III (1534–49), whom he attended as herald and historian from the late 1530s until the late 1540s. Yet though the gifts served him well, in the end they were also to be his downfall. His company and services were appreciated, but it was felt that he lacked the necessary *gravitas* to be entrusted with high office. And posterity, despite the fame which some of his works enjoyed, did not treat him very much better. One of his latter-day detractors accused him of being like a journalist, who 'can write about anything but understands nothing'.[10]

To this Giovio himself might have answered that there was one thing at least which he understood very well, and that was greatness. For all his life he hob-nobbed with the great and good; in the grounds of what was his true *paradiso*, the sumptuous villa he built on the shores of Lake Como, he erected a museum, in which he exhibited a collection of portraits of illustrious men, together with inscriptions which recorded their achievements; he wrote *Lives* and *Elogia* of famous men, in which he sung of their glorious deeds; in his *Histories* he placed history's makers and shakers at centre stage; and he did the same in his geography. In good Strabonian fashion, he held that descriptions of places should essentially be descriptions of inhabited places, and the kind of inhabitants in whom he took more interest were those who 'mattered most'. And he too favoured greatness which could abet Rome's imperial mission. Surprisingly, perhaps, that is precisely what he found in Ireland, even though he never visited the place. His *Description of Ireland* is positive like no-one else's, and he is the first Italian writer to put a proper human face and name on Ireland. He conjures up an Irish ruler, named Connatius Honel, who is every bit as perfect as Ariosto's King Oberto – although Olimpia might not necessarily have been of that opinion. Virgili certainly would not have been. For Virgili the best Ireland had to offer were some unnamed 'kinglets' who 'mostly lead life according to the manner of wild beasts'.[11]

Like Giovio, Virgili never actually visited Ireland and so also had to rely on informants for news about the place. In all other respects, though, their lives were very different. Where Giovio lived all his years in Italy, Virgili, who hailed from Urbino, studied at Padua and possibly Bologna, and took holy orders in 1496, spent most of his life as an expatriate, being sent to England in 1502 and remaining there until a few months before his death in 1555. Where Giovio aspired to success in the curial circles of Rome, Virgili found it at the English court and church, becoming adviser to kings and royal historiographer, receiving the grant of English citizenship, and enjoying preferment to a number of profitable benefices. Where Giovio stayed loyal to the pope, Virgili came to terms with the pope's nemesis, Henry VIII, and managed to hold on to his English benefices even after Henry's break with Rome and though he himself remained a Catholic until the end. Nevertheless it was as an Italian that Virgili wrote his *Anglica historia* and it was for the expertise he had gained in Italy that Henry VII sought his help to 'justify' the Tudors. When he came to England, Virgili was already famous as the author of what were soon to become two best-selling books – a collection of proverbs, titled *Adagia*, first published in 1498, and an encyclopaedia of great inventors and inventions, *On the Inventors of Things* [*De inventoribus rerum*], published in 1499 – and it was that proven ability of his to reach out to an élite readership avid to savour the fruits of knowledge produced by the new learning which brought him to the attention of England's *parvenu* monarch, who was keen to flaunt the status symbol which Italian humanists had shown no ruling family could be without, namely a modern history of its country and itself, written in the purest Latin and according to the best classical tradition.

In that Virgili served the Tudors well, and he was proud of his achievement. 'I write as an Italian', he is reputed to have said, 'and relate everything with truthfulness'. His enemies and posterity, however, have taken quite a different view. No sooner had he finished composing his *Historia*, than John Bale, the polemicist and dramatist, and later bishop of Ossory (1495–1563), launched a venomous attack on him: 'Polydorus Vergilius a collector sometime in England of the pope's Peter Pence, and afterward archdeacon of Wells, hath [...] deformed his writings greatly, polluting our English Chronicles most shamefully with his Romish lies and other Italish beggarys'. Bale, though, was not alone and the legend of Polydore the Liar was to overshadow Virgili's reputation for centuries to come. Even an Italian – the famous eighteenth-century historian of Italian literature, Girolamo Tiraboschi (1731–94) – while taking pride in the fact that 'England owed the first rounded history of the island that was ever brought to light to an Italian', felt obliged to apologize for Virgili's presumed excesses. Surely he had not 'thrown onto the fire the ancient chronicles which he had used, so that, the memory of them having been lost, his history alone would circulate in the hands of the learned'? Only more recently has the pendulum begun to swing the other way, with fellow historians coming to hail Virgili as one of the forebears of modern historiography, who wrote history as it ought to be written. 'Beginning with Polydore Vergil', it has been claimed, 'men began to look at where their information about the past came from' and to display a commendable mastery of 'the exercise of logic', as well as the ability 'to test a story by its inherent plausibility'.[12] Yet if we compare Virgili's description of Ireland with Paolo Giovio's, we are bound to conclude, at first sight, that if it was not the one whose sense of logic and of plausibility let him down, then it had to be the other.

To any reader familiar with a modern map, the most striking disagreement between Giovio and Virgili has to be their choice of simile to describe the island's shape. Likening it at first to a bird, in the printed version of his text Virgili then compared it to an egg; for Giovio it is like an oak leaf. The disagreement may do no more than reflect the standard of cartography at the time – though map-making had much improved following the rediscovery of Ptolemy, there were still many misshapen maps in circulation – but there could well be more to it than meets the eye. Ever since antiquity and until Petrarch had begun to compare the shape of the Italian peninsula to a leg and boot, the oak leaf had been the standard way of describing Italy. For a humanist it was thus a simile with privileged associations, and it is not beyond the bounds of reason to suggest that Giovio may possibly have wanted to hand down to Ireland the mantle which had once been the preserve of Italy. Certainly even what may seem as the most innocuous of details, in Giovio as in Virgili, can at times be given a much deeper significance. While disagreeing on the actual size of Ireland (for Virgili it is 'half as big as Britain', for Giovio 'less than half the size'), both concur that it enjoys an outstanding climate. But whereas Virgili uses this to underscore a failing of the Irish, for Giovio it serves as a way of wrong-footing the English, so to speak. 'It has a remarkable climate and its soil is exceedingly rich, although the Irish devote little time to agriculture', says Virgili. 'It surpasses [Britain] by the warmth of its soil, the mildness of its climate and the fairness of its weather', says Giovio, adding 'as the English readily admit'.[13]

Ireland's fauna too is cause for disagreement, as well as for a degree of what we might call political bonding. Where Virgili states that Ireland does not have any 'venomous animals' but, contrary to what some people say, that it does have bees, and that 'as pests it has the wolf and the fox' and that 'all its other animals are docile and smaller than anywhere else', Giovio notes that, 'by what is surely a miracle', it is devoid of snakes, as well as of donkeys and of 'any harmful animal, except for wolves', but that otherwise it has the same animals as 'we' do.[14] What particularly strikes Giovio, however, are Ireland's horses. 'As for horses', he says,

> the whole of the country produces pure-bred trotters which, because of their gentle gait, the English call hobbies; for the same reason they are prized by persons of tender disposition, and in France and Italy they are given as gifts to women of better families. One can see twelve such horses, of outstanding whiteness and decorated with purple cloth and silver harness, being led riderless in papal processions. They are smaller than normal, but in shape, in grace and in mettle they are quite outstanding. Ulster, for its part, produces horses which are very lively and robust, and with a strength of mind that makes them quite intrepid, as though they were born for war. After they have eaten a certain quantity of barley and fodder, their mangers are filled with burnt and crackling holly leaves, which, as the Irish have learnt from experience, is a nourishment that cures convulsions and heals the vital organs. These horses are bartered or sold to merchants from Spain and France.[15]

Thanks to its horses, at least, Ireland most definitely belongs to the same world as Giovio's.

Merchants, though, may also take some credit. Virgili makes only fleeting reference to commercial contacts between Ireland and other countries, and the only Irish goods he mentions are the pearls which had also impressed Chiericati, although as far as he is concerned they are produced not in rivers but in the 'Irish Ocean' and they are of inferior quality, being 'brownish and bluish'. Giovio too mentions the pearls, but once again they are river pearls and of a better quality, their 'weight and polish' making them almost equal 'to the pearls of the Indies'.[16] In addition Giovio draws his readers' attention to 'the huge quantities of salmon and herring, which, having been salted and smoked, are traded all over Italy', and he takes particular note of the navigability of rivers and the capacity of ports, especially the 'impetuous' Avoca [*Obocus*], 'today called the Liffey', which from Dublin to the sea 'is capable of harbouring large ships', and the 'noble Suir' which, 'augmented by two merging tributaries, provides a port in the town of

Waterford, with the safest of moorings, to the fleets of the English and the Dutch'. Among the goods which pass through these ports he lists, as exports, the furs of martens and forest cats, 'with which the clothes of the nobles are lined', bull hides, the 'woolly fleeces of sheep', and falcons, 'greatly prized for the speed of their flight and destined for the hawking of kings'; and among the imports he includes 'corselets, helmets, arrow-heads and swords from Spain and France, as well as wine, olive oil and saffron'.[17] Clearly Giovio's Ireland, contrary to what his readers might have been expecting if they followed the adventures of Orlando or were familiar with the *De Europa* of Pope Pius, was not a place from which to hurry on. Shoppers, especially those from 'better families', would have found much to keep them busy there.

Whether tourists and pilgrims would have felt the same attraction is far less certain. Virgili has no time at all for St Patrick's Purgatory and is totally dismissive of Ireland's marvels, to the point of recruiting that very work of Pius (though without acknowledgment) to justify his stance: 'as regards the miracles of Ireland, of which such a palaver is commonly made, given that, the greater the care with which they are enquired into, the further they tend to flee, we have deemed it futile to discuss them once again'.[18] On the other hand, and quite exceptionally, he actually has kind words to say about the Irish for their hospitality, recognizing that 'their guests they treat with kindness and geniality', and even going so far as to admit that the entertainment they provide is beyond compare, since they are exceptional musicians: 'they are very skilled at music; they perform with style, singing and playing the harp, but also with such force and élan, that it is remarkable how, despite the speed of their voice, tongue and fingers, they manage to keep the rhythm, which they do to perfection'.[19] Giovio also says as much, though he is more guarded in his compliment and offers it only to the Irish womenfolk. 'To those who greet them', he says of the women, 'they respond jovially, and sometimes they dance to the tune of a harp, which they hold tightly between their knees, plucking with nimble fingers at the bronze strings of unequal length, and accompanying the tunes with the playful sound of their voice'.[20] As for the miracles, he does not dismiss them, but he does betray a certain scepticism on the matter. Of the pond in which a piece of wood can change to iron and to stone he comments, as though unsure, that it is 'renowned for its wondrous nature', and of the barnacle geese that 'it seems like a miracle'. But when it comes to St Patrick's Purgatory, he is no longer willing to suspend his disbelief:

> On the other hand, regarding the oracle at the well of St Patrick where, by drawing water nine times, according to what they say, souls may be redeemed from grave sins and then, by listening to the prophecy, may have their future foretold, we cannot at present say anything definite, unless this burrow, dug by the enterprising ancients in a hidden cliff on the furthest shore of the island, where it is exposed to the blasts of the north-west wind and where a spring of fresh water rises, is to be considered a miracle. For our part, we will leave such fable-like matters to the inquisitive superstition and the prayers of the pilgrims.[21]

Quite why the Purgatory has been transformed so radically in so short a time is somewhat of a mystery, not to say a miracle. What we can tell for sure is that Giovio does not appear to have shared his views on the matter with Chiericati, any more than he was willing to share Ariosto's belief in an *Ibernia* that might be *fabulosa*. Not on that count at least.

Confused – and confusing – though he may be about the nature and location of St Patrick's Purgatory, Giovio is on much safer ground when it comes to carrying out the geographer's prime task of partitioning the land. Of the 'many regions' into which the Irish are divided, he logs the four 'most famous' – Ulster [*Holtonia*], Leinster [*Laginia*], Connaught [*Connatia*] and Munster [*Momonia*] – together with the four rivers which separate them—the Liffey/Avoca [*Lipheus/Obocus*], the Boyne [*Boandus* or *Buvinda*], the Shannon [*Sineus*] and the Suir [*Suirus*, once called *Modanus*] – and the 'four holy metropoles' of Armagh [*Armacana*], Dublin [*Dublinensis*], Cashel [*Cassilensis*] and Tuam [*Tuomensis*], besides the towns of Limerick [*Limericum*], Waterford [*Vaterfordia*], and Dublin, 'the largest of all the cities', which 'has grown remarkably thanks to English trade'.[22] Virgili for his part has more of a struggle to get the divisions right. Mentioning only three of the rivers (Boyne, Suir, Shannon), in the manuscript version of his description he takes the ecclesiastical provinces as his standard and makes no reference whatsoever to the lay provinces, but in the printed version he does the very opposite. There only the lay provinces are recorded and all references to the ecclesiastical ones have been removed. At the same time Virgili proceeds to downgrade all the island's centres of habitation. In the manuscript he lists eight or nine of them, six of which are called 'cities' [*civitates*], one a 'borough' [*oppidum*] and one a 'port' [*portum*]. In the first edition they have risen to twenty, but the scales are tipped in the opposite direction, with fourteen called 'boroughs', one a 'market town' [*emporium*], and still only six named as 'cities'. And by the time of the third edition there are no cities left, all six having now been recast as towns [*urbes*].

To have the Irish live in towns was itself possibly a step too far, since *urbs* implied a degree of sophistication in building and planning, and the best the Irish could manage were 'houses of either stone or wood, with walls made of mud'. More importantly, as Gerald of Wales had already noted, the Irish were essentially 'men of the woods', or better, 'backwoodsmen' [*silvestres*], who made a point of rejecting urbanity, and even more so, of rejecting civility, which was the very hallmark of the *civitas*. As Virgili states unambiguously, they were 'wild, uncivilized, stupid and hostile, and because of their lack of civilization and their coarse manners, they are called backwoodsmen'. An Irish city would thus have been an oxymoron. Nevertheless, and tellingly, as he wrote cities out of the Irish landscape, Virgili did leave in one mention of the word. 'It is the cities which are least uncivilized', it says in the printed version, 'for they obey the English king, wherefore they absorb more honourable manners'. Irish cities, that is, are virtual cities: would-be cities which exist only by virtue of verbal gymnastics, requiring both an oxymoron and a litotes to be perceived as such. They are cities only to the extent that they are English, and they cannot claim to be civilized, but only 'least *un*civilized'.[23]

Giovio, for his part, takes the opposite view. True, there are a number of cities in the Ireland he describes and he too acknowledges that these have 'grown remarkably' thanks to the presence of the English, but the fact that the Irish mostly turn their back on city living he sees as a virtue. There is foresight in their choice of habitation – 'they live in villages, and when choosing a location they take into consideration the advantages offered by marshes and mountains' – and at their best they shun even village living, as demonstrated by the case of the greatest of them all, Connatius Honel. Connatius, we are told, 'as behoves an intrepid and alert warrior, avoids the villages and spends all his time, summer and winter, in camps. In true military fashion he seeks shelter from the rain storms in bowers covered by foliage, under tents of hide or canvas'. Indeed, it is through his very rejection of what constituted the main buttress of a city that Connatius proves his greatness:

> Honel's royal seat is called Armagh and, because of its particular authority in religious matters, it holds the primacy of all Ireland; but it is not surrounded by any ramparts, and so too are the other strongholds of Ulster exposed, through lack of walls, since he considers it more noble to defend his state from its enemies with his bare hands and in open battle than to protect it with fortifications and garrisons.

Not for Giovio, in other words, Chiericati's 'lords, who are little more honourable than our peasants'.[24]

The 'religious matters' which lend authority to Honel's royal seat are also cause for disagreement between Giovio and Virgili. It is clear that, being backwoodsmen, Virgili's wild Irish could not be paragons of Christian virtue, yet as Chiericati had been forced to recognize and Virgili himself could not but acknowledge, their devotion to the pope was quite exceptional. It was thus impossible to ignore it, and even less so as it provided the premise upon which Virgili built his narrative and his description of Ireland. Having arrived in Ireland, Henry II, we are told, discussed with the Irish 'kinglets and bishops' he had summoned the question of 'how to render dominion over the island permanent for himself and his heirs'. But this, the Irish answered

> could be done only with the authority of the Roman pontiff, since from the very beginning, after the Christian religion had been received, they had surrendered themselves and all their belongings unto his power; and they kept repeating that they had no other master except for the pope himself, which is something they claim even now.[25]

Coupled with such devotion, however, was a tendency among the Irish to practice religion in a manner which today is commonly referred to as *à la carte*, and it was on account of that tendency that the pope had made up his mind – 'with little hesitation' – to 'join Ireland to the kingdom of England', having concluded that, since there were 'many things that pertain to our religion' which the Irish did not 'yet observe', they would 'become more civilized and gain a better understanding of divine matters, if they all at least obeyed one, Christian and most mighty king'. Yet even the pope, it would seem, did not wish to overstep the mark. The first of the 'many things' the Irish did not 'yet observe', according to the manuscript version of Virgili's description, was 'holy baptism'. But in the printed version this has been replaced by tax avoidance, with the pope now said to have 'received no income from there'. The Irish, in short, had to be both good and bad Christians, and so it is little wonder that it is an oxymoron once again which serves to encapsulate them: 'they are called backwoodsmen, because they mostly lead life according to the manner of wild beasts, although, in their beastliness, they practice the Christian religion chastely'. The Irish practice beastly chastity. Ariosto certainly would have agreed with that.[26]

A similar paradox lies at the heart of Virgili's assessment of how the Irish handled education, a matter of no small concern to humanists, who were wedded to the belief that education makes the man (and occasionally the woman). As we saw in an earlier chapter, it was thanks to his and their mastery of Latin that Guerrino il Meschino found his way among the Irish, and sure enough we notice that Virgili gives full credit to the Irish for their knowledge of the language and the benefits they apparently derive from it. 'They are not without learning', he comments, 'for they study Latin from childhood, as well as canon and civil law', continuing: 'moreover they are sober and show great endurance of hunger'. Yet in the sentence just preceding this one he had cast serious doubt on their linguistic competence and had drawn the same conclusion about the link between their language and their character, but in reverse: 'they all speak the same language, which they pronounce as though they were babbling and groaning; they are quick-tempered, savage and vengeful; and in addition they are not to be trusted but take pleasure in lying'. Studying Latin, it is clear, was no cure for backwoodishness, and in order to be sure the point would be well made Virgili added 'as though they were babbling and groaning' on mature reflection, in the printed version of the work. A similar kind of point is made about funerals in Ireland, which, contradicting his opinion about Irish musicianship, Virgili dismisses as a lot of fury signifying nothing: 'they lament the deceased at length and with great clamour; and if they are important men, they hire women to wail by the body'.[27]

For Giovio, on the other hand, funerals, with which he ends his description, are one of several signs of the noteworthiness of Irish customs, and what is particularly noteworthy about his own description of them is that it shows an ability to appreciate such customs on their own merits:

> The funerals of those who are killed are not conducted with cries and tears, but with songs of praise; and the relatives hire wailing women, so that their laments may add splendour to the ceremony; indeed they consider it a matter of pride for the family not to mourn overly those who have met a great fate in death. It is more dignified, they hold, to venerate one's beloved not with tears but through lasting memory, for it would be shameful to begrudge them an honourable exit from life and the incomparable gift of final peace.[28]

We find the same openness of mind, not to say sympathy, in Giovio's assessment of the religion and education of the Irish. He, for one, has no need of oxymorons. There are no *silvestres* people in Giovio's Ireland, only *silvestres* places, but even there religion is respected: 'all the Irish, even in harsh and wooded places, observe the Christian rites'. What is more, Giovio actually shows admiration for a church where keys do not get lost at the bottom of St Patrick's well: 'they are taught by teachers appointed by the bishops for their suitability as masters of Latin and music; and on account of their formation and learning, these men are promoted to wealthy benefices'. One could almost imagine Giovio saying: 'Go there, Paolo, what are you waiting for?' Giovio also seems to admire the positive effects which the Church apparently has on Ireland's legal system: 'it is not their custom to prolong their lawsuits, as they have no place for advocates, and disputes are ended by summary judgement or arms; only priests attend the tribunals of the bishops, since they obey the decrees of pontifical law'.[29] It is clear why Giovio, unlike Virgili, had no need to do away with Ireland's ecclesiastical provinces. Virgili required an Ireland which called out to be settled and righted. Giovio's was properly ordered.

The one aspect of the Irish understanding and practice of Christianity with which Giovio had some difficulty coming to terms regarded their sexual mores. Virgili had also found the Irish wanting in that respect, and what is more he had made of that particular failing of theirs one of the prime considerations in the pope's decision to subject the Irish to the king of England: being 'uncouth and backwoodish', they 'did not yet observe matrimony in the proper way (each, indeed, had as many wives as his wealth would allow)' and therefore needed to obey 'one, Christian and most mighty king'. Giovio is more explicit on the matter of Irish promiscuity, but once again he attempts to understand it from their own point of view, though not without first having laid the blame for it on the women of Ireland, of whom he writes:

> Every one of them is of doubtful modesty, and slight indeed is their reputation for chastity, since whole families sleep together in very large beds, where, in darkness, they recklessly set shame aside. The kinglets and the nobles maintain both a wife and mistresses, and it is not considered scandalous to repudiate one's wife and, having returned to her her women's jewels, to slip the same wedding ring on to the finger of a concubine. The others too freely engage in sex, and they do not number lust amongst the serious crimes forbidden by Christian laws, for guilelessness alone is the gauge by which they earn a name for honesty.[30]

'Christian laws' notwithstanding, the practice may not be without some merit – one senses Giovio thinking – and the name not without justification.

The Irish sense of dress also meets with Giovio's approval. He draws attention to the 'patterned border and fringe' which, 'for the sake of elegance', decorates the woollen cloak which almost all the Irish wear, and he notes how the borders of their saffron tunics are 'skilfully embroidered with silken threads of different colours'. For Virgili the cloaks are 'shaggy' and saffron is a sign of the unusual sense of hygiene of the Irish, who use it as a form of what we might call dye-cleaning: 'they wear a tunic made of linen cloth, which they do not change before it is worn and which they dye with saffron, so that it will not get dirty'. Virgili also highlights the peculiarity of the Irish head- and footwear, or lack thereof, as though to underline once again the half-way status of the Irish between man and beast. Only the nobles, he informs us, cover their heads and wear shoes. The rest of the population, going barefoot, have developed skin which is so tough 'that children, whose feet have not yet been hardened by long use or work, can run with remarkable agility, even on rugged ground'. To Giovio on the other hand sartorial differences serve to underscore the normality of the Irish, especially in relation to Italians. He tells us that, for their woollen cloaks, princes prefer to use scarlet, 'leaving dark blue and red for the common soldiers', and that 'it is for this reason that we here call Hibernian, as coming from there, the cloak ordinarily worn by the captains of ships and their officers'.[31] Cloaks, horses, fish and oak-leaves: the Italians and the Irish certainly had much in common.

Only in the way the Irish wear their hair does Giovio find a hint of the exotic. Young girls, he informs us, 'display theirs gathered in small knots', but older women cover it and adorn it 'with strips of saffron cloth, which they wrap around it many times, in the manner of Turkish women'. Chiericati, we may remember, had said that they cover it 'after the fashion of Carmagnola', that is, after the fashion of a town in Italy. As for the men, Giovio writes, possibly with a touch of bemusement, that 'so as to appear more ferocious in battle, they march in rough and ready fashion with their hair unkempt and cut above the ears, often with their beards shaved off, and with their upper lip bristling terrifyingly with a curled moustache'. The bemusement, though, is mild compared with the mocking tone of Virgili, who once again disagrees with Giovio on actual matters of fact, stating that

> they do not cut their hair above their ears, although some, in the old fashion, cut all of it off at the back of their head and wear it long in front. They frequently shave their beard, leaving only their top lip unshaven in order to look terrifying, or so we may reasonably believe.[32]

The Irish, there is no doubt, make it very difficult for Virgili to suspend his disbelief.

A further challenge to his credulity is provided by the Irish attitude to work. Both he and Giovio note that the Irish prefer to hold back from work, but whereas Giovio, once again, tries to interpret this in a positive way, Virgili is hard put not to call it downright laziness. 'Among their pleasures', he insists, crafting his sentence in such a way as to foreground their witting consensus, 'is holding back from work, and they do not practice agriculture to any great extent, nor any of the other crafts, except for music'. Giovio too is of the view that the matter deserves to be emphasized through foregrounding, and it is thus the very first trait he mentions in his portrait of the Irish, but as far as he is concerned, it is for them a matter of principled consideration, which forms the very basis of the moral code that makes them into 'a truly simple people, ignorant of luxury and unspoiled by temptations from abroad': they 'hold the plough in great contempt, shun the hoe and avoid all toil save war, which is well nigh the only thing, besides honour, which they deem worthy of pursuing'. Nevertheless, and remarkably 'unspoiled' though they may be, Giovio does have some difficulty in coming to terms with what could also be interpreted as their stubbornness, and for once he almost seems inclined to take the side of the English. 'What is more', he adds,

> their stubborn minds have decreed that the wild metal-bearing mountains should be spared, and they refuse to allow the removal of gold or silver extracted from quarries or mines through the zeal and hard labour of the English, for they consider such things to be exceedingly harmful to one's life and character.[33]

It is not always possible for 'one' to be at one with the Other.

As the second marker of the 'truly simple' character of the Irish Giovio focuses on their diet, and this provides what is perhaps the most curious and revealing case of disagreement between his description and Virgili's. According to Giovio, being

> happy enough to feed on meat, milk, cheese and honey, besides fish and the catch from fowling and hunting, they do not covet the gifts of Ceres and of Bacchus which are served at the tables of their princes, for they ordinarily prefer cakes made of common wheat and beer served in cups, and they feast on great quantities of pulses.

Virgili, in the manuscript version of his description, at first makes the Irish into vegetarians: 'as a very hardy breed of men', he writes 'they go without milk and feed on fruit, and when they eat bread they make it from oatmeal'. In the published version, however, he has a change of mind: 'being an exceptionally hardy breed of men', he now asserts, 'they do not live off grain much, but for the most part off milk and meat; when they eat bread, they make it from oatmeal'. It would seem reasonable to assume, since that is what he has been given credit for by his admirers, that the change of mind was the result of second thoughts on Virgili's part about the reliability of the source 'his information came from'. But as it so happens, it was other instincts which brought about the change. Virgili's description of the diet of the Irish in the printed version of the *Anglica historia* is a word-for-word reproduction of Julius Caesar's description of the diet of the Gauls in *The Gallic War* [*De bello gallico*]. 'Neither do they live off grain much, but for the most part off meat and milk', says Caesar of the enemies of the Romans.[34] The borrowing – and it is by no means an isolated case in his description of the Irish – speaks volumes not only for Virgili's method, but also and above all for his intentions.

As was required by convention and was expected by those who made 'gifts' to his 'honoured hands', Virgili's description of Ireland took as its starting point that it mattered just as much where the information went to as where it came from. The *Anglica historia* had been commissioned by Henry VII and was dedicated to Henry VIII, and in the dedication Virgili makes no secret of his belief that history must side with those *for* whom it is written. 'You above all', he flatters Henry

> know perfectly how to pass correct judgement both on deeds done and on deeds to be done, you who, born to empire [*imperium*] and justice, are gifted with great prudence and, exceptionally learned in divine as well as human letters, excel in all good arts. And so by your singular virtue you easily surpass not only the glory of those princes who are alive today but also the memory of all of antiquity.[35]

In other words, only a king with 'empire' and 'justice' on his side has the power to 'correct' the course of history, that is, to make sure that the past ('deeds done') lives again in the future ('deeds to be done'); and this in turn means that the historiographer has the duty to craft the past in such a way that it can have the future it deserves, a future that will allow 'those princes who are alive today' to be present in tomorrow's 'memory' as singularly as 'all of antiquity' is present in today's (while ensuring that their enemies all eat in the same way).

This aligning of past and present, in what amounts to an immobilizing of the course of history, is achieved most notably by Virgili, as far as Ireland is concerned, in the setting of the description. The description is set in the chapter which recounts the initial attempts by the English to conquer Ireland, at the time of Henry II, but it describes the Irish as they are 'now', at the time of the early Tudors. And to drive the message home, the postscript to the description, in the chapter on Henry VII, in effect tells the readers that if they want to find out what the Irish are like *now* (and by implication forever will be like), they should go back to the chapter on

Henry II to find out what they were like *then*, for nothing has changed: 'in the whole of Ireland there are two types of men, as we illustrated in the appropriate place above, in Book Thirteen of this work ....'[36] The place is 'appropriate' because it is the job of the historiographer to make the repeatability of history predictable. But it is also his job to make it appear as the manifestation of justice; and he can only do that by demonizing those who, in the estimation of anyone 'learned' and 'virtuous' enough to 'pass correct judgement', cannot but be at the receiving end of justice. Justice, as we have seen, is the preserve of the kings who are born to it, but the Irish have no kings, only 'kinglets', and the one such kinglet who stands out above the rest Virgili duly cuts down to size. The part of the island which is not under the English, he had said in the manuscript, 'is ruled by several kinglets, whose leader *is* the king of Connaught'; but in the published version that leader can no more than '*call himself* the king of Connaught'. And to underscore the point, Virgili then makes Ireland's kinglessness into the effect and cause of its seditiousness. 'They have several kinglets, to whom they are subject and who continuously fight among themselves', says the manuscript of the 'backwoods' Irish, and 'for this reason they are more ferocious than the rest of the Irish, and being more eager for new things, they are considered more disposed to political change'. The printed version specifies: 'They have a great many kinglets, who continuously fight among themselves, wherefore they are more ferocious than the rest of the Irish and, being extremely eager for new things, they resort to plunder and banditry, liking nothing so much as insurrection'.[37]

'New things' are what the Irish need of course, but clearly these cannot be the kind of things brought about by 'insurrection', and that too is what Virgili's description of Ireland serves to illustrate. How the 'things' will pan out is made manifest by the manner in which the Irish fight. Having started out, in the manuscript, from the same premise which had allowed Gerald of Wales to predict the victory of 'our western Alexander', namely that the Irish fight naked, or near enough, Virgili is then forced, in the published version, to recognize that they are 'now' (that is, in the early sixteenth century) learning to adapt their methods, and therefore the manuscript observation that they 'fight without body armour, and they mostly ride without pillions' is modified and expanded to read as follows:

> They fight without body armour, which they deem equally conducive to audacity and physical vigour; and bearing arms they consider a burden although, gradually growing wiser by being exposed to danger, they are just *now* beginning to do so. They ride without saddles, although they do *now* use tiny pillions, but without any tackle.[38]

However, lest the judges of deeds done and to-be-done draw the wrong conclusions about the audacity and vigour of the Irish, and their capacity to fight back and defend themselves, and thus to thwart the progress of empire and justice, the postcript makes it very clear that in fact the Irish have little chance of success. Commenting on the story of Perkin Warbeck (c.1474–99), the false pretender to the throne of England who in 1495 had come to seek the help of the Irish, Virgili had written in the manuscript version: 'having stayed there a few days, he understood that the Irish, who do not cover their bodies with any armour when fighting, are worth little in battle *outside their native land*'. In the printed version on the other hand he decided to write: 'he stayed only a few days, for he understood he had little chance of victory by relying on the Irish alone, since they did not cover their bodies with any armour when fighting and therefore were worth little in battle *against the English*'.[39] The English, in other words, had nothing to fear. However much the Irish might have learnt from experience, Ireland was England's for the taking.

In Giovio what experience teaches is the very opposite. 'There once was a time', Giovio writes

> when the English, having roved about the island with large armies, aspired to the lordship of the whole of Ireland; but when the Irish, who had long been caught up in internecine warfare and had paid dearly for such madness, awoke to the danger and focused their minds once again on public safety, they joined forces and broke the strength and the spirit of the English to such an extent, that they left them with only a tiny part of the island, which, as a matter of fact, they were generous to concede to them; for in the rising against the English Ireland armed more than 40,000 foot-soldiers and horsemen.[40]

Not only could the Irish fight, that is, but they were driven by a sense of common purpose, which meant that the English had everything to fear regarding their kings' chances of 'rendering their dominion over the island permanent' (to quote the words of Virgili). What is more, it was precisely because they themselves had no kings that the Irish posed such a threat – and had every right to. The Irish, Giovio writes,

> owe equal allegiance to their various kinglets, and they have never allowed all power [*imperium*] to fall into the hands of a single person, for that, they think, would deprive them of their liberty. Such is the disposition of these wild men that they rejoice greatly in the name of justice, and it is their greatest concern that no one should command such forces and consequently, as is often the case, become so overbearing and despotic as to subdue the others and seize control of the island.[41]

Giovio could not have been more explicit: empire and justice are now in the gift of the Irish.

Yet, concerned though the Irish were that no single person should have a monopoly of *imperium*, there was in fact one such person in their midst, and that person was Giovio's great Irishman (his King Oberto, so to speak): Connatius Honel, or in other words, Conn Bacach O'Neill (1480–1559), the Ulster Gaelic 'kinglet' who was the chief opponent of the English and was to remain a thorn in their side until early in the reign of Elizabeth I (1533–58–1603), even though he had surrendered to Henry VIII and been regranted his lands as the first Earl of Tyrone in 1542. 'By his great authority and military achievements, as well as his ancient nobility', Giovio writes in the sentence immediately following the one just quoted, '*Connatius Honel*, the Prince of Ulster, surpasses all others. He is constantly at war, now against the neighbouring English, and now against the kinglets who from all sides press in upon the borders of Ulster'.[42] Conn's way of fighting, moreover, is not the way of the seditious, who fight to destroy or merely for the sake of fighting, but it is purposeful. There is a purpose, as we have already seen, to Armagh, his 'royal seat', being without ramparts, and there is a purpose to his wish to spend all his time in camps. That purpose, we then learn, is the realization of the natural impulse of a society which holds its warriors in high esteem, warriors who most certainly do not fight naked, or recklessly:

> Soldiers are held in great esteem, the first place going to horsemen and the second to those heavily armed foot-soldiers who are like the veterans of old, fighting war upon war and brought up in camps, and who in their own language are called galloglasses. Conn O'Neill stands out amongst all for the might of his cavalry and he

> alone maintains heavily armed foot-soldiers; he shares land and fortune with them, rewarding them, for want of money, with food, and also clothes, arms and horses. The horsemen wear a corselet and a helmet, in their left hand they hold a Spanish lance together with the reins, while their right hand propels sling-strapped javelins with remarkable precision. They usually keep several of these either packed in their sword-belt or pressed against their thigh; others are handed to them by their foot-servants. Thus from a distance they fight with javelins, but close up, having transferred the lance with great speed from their left to their right hand and raising it aloft, as they hold it towards the middle, they fight by striking with it repeatedly. They bring along two horses each, one to carry them outside the battlefield, and the other indeed, which goes empty-saddled and is an excellent fighter, to transfer to at the very moment they are called to arms by the sound of the trumpet. It is not the blare of the trumpet, though, that rallies their fighting spirit but the wail of a mighty wooden horn. Such is the horseman's dexterity in restraining and spurring his mount, that by ducking and dodging he avoids enemy darts with great agility, and he snatches any that are scattered on the ground in his outstretched left hand, hanging from the saddle by the strength of his knees. The veteran foot-soldiers, for their part, besides their helmet and their corselet, protect their throat with a large band of chainmail; and after they have thrown their javelin, at the beginning of the battle, they then use a German axe. When they have been formed into a steady line, light troops rush forward from the wings, to the sound of the bagpipe, and fight with darts and with arrows, which they shoot from short wooden bows. The galloglass fights with such valour, that he thinks only of conquering bravely or dying honourably where he stands, with no fear of death.[43]

Giovio could not have been more complimentary. So valorous are the Irish soldiers that they are like the Roman soldiers 'of old'. A *galloglatha* is a *veteranus*. The Roman 'done', to be sure, bodes well for the Irish 'to-be-done'. And indeed, it bodes the very opposite for the 'to-be-done' of England.

The one Irish miracle Giovio seems prepared to credit is the one performed, or almost, by the soldiers of Agricola and recounted by the very author (Tacitus), as we saw in Chapter One, who had wondered whether there was any 'use' in the friendship of an Irish kinglet:

> Two coastal regions in Leinster, called Meath [*Midia*] and Fingall [*Fingallia*], are held by the power [*imperium*] and the garrisons of the kings of England. From there the crossing to England is shortest, with the island of *Mona* in between, to which, according to Tacitus, Roman legions managed by some miracle to swim across, having discovered shallow fords and benefited from the tide of the ebbing and flowing ocean, so that from *Mona* [...] they ambitiously planned to invade Ireland, in the hope of a

> triumph; and Julius Agricola believed it could be subjected with the help of a single legion and few auxiliary troops, and easily kept in peace.[44]

Put otherwise, what is truly *fabulosa* about *Ibernia* is the ambition of those who believe they can 'easily' visit their *imperium* upon it, unheeding of the valour of Irish soldiers and – for Ireland, it would seem, cannot make sense without an oxymoron – of the valour of their mightiest 'kinglet'.

The greatest merit, though, of Irish soldiers and their leader is not so much that they know how to fight, but that they know when not to. They have, as we have seen, a sense of the common good. 'Public safety' can dictate to them when to leave their 'internecine' squabbling to one side and 'join forces' so as to 'break the strength and the spirit' of their common enemy. Better still, it can tell them when to be magnanimous, showing generosity to the English by allowing them to remain in control of 'a tiny part of the island'. And it is in that spirit that Conn O'Neill finally decided to make peace with the English:

> Recently Conn O'Neill himself, following a rigorous muster among his fellow Ulstermen, led 4,000 horsemen and three times as many foot-soldiers into battle; but he was finally won over through the efforts and by the gifts of Thomas Howard, the famous victor of the Scots, and he made peace with King Henry.[45]

True, it could seem as though it was venality instead of 'public safety' which motivated Conn; but the very last words of Giovio's description can leave us in no doubt. By burying the dead the way they do, the Irish make it clear for all to see that they understand the importance of 'an honourable exit' and of 'the incomparable gift of final peace'. Virgili, for his part, chooses to end on quite a different note. His own very last words appear almost dismissive and seem to suggest that, like Pope Pius and Orlando, not to mention Strabo, he was in a hurry to move on. Having brushed aside Ireland's miracles, 'of which such a palaver is commonly made', he concludes: 'we believe that we have said all that needs to be said here about this island. So let us now return whence we digressed'.[46] Yet though to us admitting to have digressed reads like an admission of failure, or at best an apology, to Virgili it was in fact a way of showing off. It demonstrated that he knew the rules.

The rules, as laid down by the ancients, stipulated that chorography was to be an inseparable companion of history; and since the ancients said so, humanists said so too. The point had first been made by Cicero, in his dialogue *Orator*, a manual on the art of rhetoric, but more recently it had been restated by Giovanni Pontano (1422–1503), the poet and prime minister of Naples, in his dialogue *Actius*, a trend-setting primer of humanist style and genre, written in about 1499. Virgili too had said as much, in *The Inventors of Things*: 'the edifice of history', he had stated there, 'demands chronology as much as chorography'.[47] The reason history demanded such a combination was that chronology, on its own, threatened to negate its repeatability. The natural tense of chronology was the past historic and its main focus was that which in the course of time had been unique. The natural tense of chorography, on the other hand, was the present, or better, the present continuous, and its focus was what remained unchanged – because the world and human nature were unchanging. Thus chorography was in a sense even more important than chronology, for only chorography could anchor the present firmly to the past, and so to the future. However, chorography also served an aesthetic function. Chorographical descriptions, as Pontano had explained, were best supplied by way of 'digressions'; digressions were an indispensible way of ensuring 'variety'; and it was variety which gave the readers pleasure (Horace's *dulce*) and so kept them firmly glued to the page. That is what Virgili's description of Ireland was meant to do.

At the same time it was meant to do something far more significant, by attending not only to the readers' need for *dulce* but also and above all to their need for *utile*. As Pius had explained in *De Asia* – which, together with *De Europa*, was constructed according to those very same principles – chorography had the practical aim of setting before the readers' eyes the 'field' in which historical events unfolded, so as to enable them to visualize the past, as it were. More importantly chorography, in its alliance with a sense of history which gave primacy to war, and thus to winners and to losers, served an ideological purpose. 'Crispus – Pontano had written, referring to the *War of Jugurtha* [*Bellum jugurthinum*] by Sallust (Gaius Sallustius Crispus: 86–35 BC) – describes the lay of Africa, so that, for sure, both the land itself be known, and who one's friends and who one's enemies in every province'.[48] Chorography, that is to say, was intended to be partial. It had to take sides, so that, while chronology reported what happened on the battlefield, the prudent and the learned readers could form a picture in their mind of who the friend was and who the foe of empire and justice. For chorography too, in other words, there had to be winners and losers. That, of course, had also been the point of view of Strabo: geography (cosmography, chorography) was all about the where and the who that mattered most to those who mattered most, and the where and the who that to those same people mattered least. There was a geography of winners and a geography of losers, and above all geography was written for the benefit of winners. Virgili, as he described the Irish in his history of England, bore all of this in mind.

Giovio, for his part, experimented with a different approach. He too was aware of the rules which stipulated that some peoples should be more equal than others, but he explicitly distanced himself from them. In the dedication of the *Descriptio* (which, as has been pointed out, was intended as the first part of a description of the whole world) he wrote: 'following the example of the ancients, we will describe Europe first, beginning with Britain, even though by absolute right we should begin with Italy, once upon a time the victor of all peoples, and with that city ever destined to be great by virtue of its empire'. And just as he made 'the victor of all peoples' take a back seat, he did the same with chronology, thus bringing about the separation of choro-cosmography from history. In the dedication he also says:

> Having put aside our *Histories,* and with our mind set on entertainment, we will leave the way of ancient studies, as is only right, and, in a more relaxed style, we will take up a most enjoyable kind of writing, so that, with the novelty of an uncommonly pleasant and beautiful type of work, we can bring pleasure not only to the learned but also to those of tender disposition and those who lazily enjoy leisure [*otium*].[49]

The *Descriptio*, in other words, is a form of anti-history, the point of which is primarily to satisfy the need for *dulce* of its readers. Indeed, to such an extent does Giovio set history aside in the *Descriptio* that the work was deliberately published well after its sell-by date: it came out approximately one year after the death of Henry VIII, but the description of England ends with Henry's marriage to Catherine Parr in 1543. More significantly, even though England's divorce from Rome had by then been fully consummated and the country had come under the rule of the fully Protestant Edward VI (1537–47–53), what underlies the description of England, which constitutes the main section of the work, is the hope and belief that Henry will one day be cured of the madness which had led him to break with Rome. Unless this was an oversight due, say, to pressure from the publisher, we are bound to conclude that it was intended as a deliberate signal to the readers and a confirmation of the message of the dedication. Put history aside – it warned the readers – this work is of a different kind.

If we were to judge from the description of Ireland alone, we would say that, since the work is not history, it has to be propaganda. Not only is it sympathetic towards the Irish and particularly towards Conn O'Neill, to whom the Catholic and imperial powers had sent a delegation in 1542, in the hope that he might be another King Oberto, so to speak, but it is the work of a man who, besides being a strong believer in the supremacy of Rome, was also a close ally of Henry VIII's Catholic cousin, Cardinal Reginald Pole (1500–58), a one-time claimant to the English throne and a soon-to-be frontrunner in the election to the papacy, who is in fact made to figure prominently in Giovio's description of England. Unfortunately, such an interpretation does not really stand up to scrutiny, for all that Giovio himself had apparently not given up the hope that Pole might still succeed to the throne of England.[50] By 1548 the expectation that Conn could be a willing and effective ally of pope and emperor had turned out to be ill-founded and the hope was no longer entertained by anyone in Rome, it would seem. More importantly, Giovio has nothing whatsoever to say on the matter in the works where one would have expected him to discuss it in greater (historical) detail. Both in his *Histories* and in his letters Conn and the Irish are conspicuous by their absence. More importantly still, though Giovio may have admired Ireland and the Irish, he admired England and the English even more.

The greatest compliment he pays to the English, in the description of England, is to recognize them as the torch-bearers of learning, thanks to whom Europe, and above all Italy, emerged unscathed from the darkness of the Middle Ages:

> there is nothing more honourable, more godly and more sober than the education and the instruction which the young receive there, and it is therefore no surprise that that is where the minds came from which filled the whole of Europe with admiration

> at the subtlety of their dialectics, of their philosophy and also of their theology, especially when our ancestors, having repudiated the best of ancient authorities on intellectual matters, prized nothing more highly than the fallacies of entangled arguments and the convoluted sophistry of disputations.

Put otherwise, the Renaissance was the fruit of English 'godliness' and 'honour'. Coming from an Italian humanist, that was no mean compliment. True, the English also had their defects, not least their obsession with 'luxury and the endless pleasures of the palate', and naturally there was the serious matter of their 'iniquitous impiety' and the 'execrable temerity' with which they undermined 'the cause of religion'. But in fact it was not they so much who were to blame for such impiety, as it was their wayward king. And their wayward king, when all was said and done, was not so much an apostate as he was a tragic madman: a fallen hero led astray by an evil counsellor (his Chancellor, Thomas Cromwell) and by the 'wicked and ambitious designs' of a seductress who preyed upon his 'uncontrollable lust' (his second wife, Anne Boleyn). It made sense therefore to imagine that such a man might one day be cured of his madness. And that is precisely what Giovio does:

> Although, with the blood of the most illustrious and most innocent of men Henry has soiled so many marriages, the fruit of his uncontrollable lust, and although we see no end to that so far, nevertheless I must believe that all will turn out happily for him, should both piety and self-restraint move him to take a draught, as he would of a providential medicine, from that most salutary of books which, most lovingly and piously, his relative, Cardinal Reginald Pole, has dedicated to him.[51]

Historically speaking, of course, there was no point whatsoever in hoping for a happy end of the sort. In 1548 Henry was dead and buried. But setting history aside and speaking metaphorically, there was no reason not to do so, for it was never too late for a 'providential medicine', especially if that medicine came in the form of a book. Specifically, what Giovio had in mind was the *Defence of the Unity of the Church* [*Pro ecclesiasticae unitatis defensione*], written by Pole in 1536 to demonstrate the errors of Henry's actions. But might he also have been thinking of his own *Description of Britain, Scotland, Ireland and the Orkneys*?

What is for sure is that the work lends itself to being read as a four-act morality play, in which virtues combat vices so as to heal the deadly wounds of mortals. In Act One (England), we are witness to the tragic downfall of the heroic king of a noble land, and the act concludes – which it could not have done if Henry had been shown to be already dead – with a prayer for his redemption. In Act Two (Scotland), we are made aware of how, in practical terms, the redemption could come about – through the 'very old tradition' of getting the Scots to launch a 'hostile attack' upon the English with the assistance of a Continental power – but at the same time we are made to realize that any such attempt could well be futile, for the characters of this act, like the characters of the previous one, are also locked in errors and delusions: their history is a history of constant civil wars and 'tyrannies', their 'ambition and lust for power' is insatiable, and their love of food more insatiable still, since 'by the abundance of their dishes and their gluttony, which to our people seems almost incredible, they surpass the English'.[52] However, in Act Three (Ireland) we are lead into a different world, where we gaze upon the real world's upturned image and glimpse the promise of a better and redeeming Otherness.

The Irish, we find out, are 'happy enough' with their lot, and they stand unmoved by the promises of evil. A 'truly simple people', they are not only 'ignorant of luxury' and 'unspoiled by temptations from abroad', but they do not even 'covet the gifts of Ceres and of Bacchus', and their leader, in the name of liberty and justice, refuses to be seduced by the lure of tyranny or empire. Better still, where Henry knows only 'wicked greed', Conn is unperturbed by 'want of money' and even has the 'generosity' to share his land with those who come to steal it. And where Henry thinks only of his own desires and cannot tell what is good for his country, nor when enough is enough, Conn and the Irish are able to set aside civil discord for the sake of peace and the good of all. They know better too than to split from Rome, and their religion is a model of orthodoxy. It is all englobing in its Romanness – being manifest not only in the island's 'holy metropoles', in its bishoprics, its law courts, and its schools, but also in its 'wooded places' and its promotions, and even in its horses – and it has no pretensions beyond the simple observance of 'the Christian rites'.

Admittedly, in an earlier incarnation Henry had himself been a model of orthodoxy, when, 'thanks to the remarkable fertility of his mind he had published a book against the arch-heretic Luther', but such fertility – we stand warned by the darkness of the Middle Ages – could also lead to 'convoluted sophistry', and what is for sure is that it did not prevent Henry's descent into madness. What might well have done, however, are the marital customs of the Irish which, in view of Henry's own marital woes, come across as eminently sensible. If, like his Irish counterparts, Henry had been in a position simply to 'repudiate his wife' and 'slip the same wedding ring on to the finger of his concubine', without having to promulgate, as he did, 'iniquitous' and 'impious' laws that had no regard for 'ancient treaties, left inviolate by so many earlier kings', how differently, and so much better, everything would eventually have turned out. Un-Christian though the customs are, they would nevertheless have guaranteed that Henry and the English remained Christians true to Rome. And even the summary justice of the Irish, not to mention their obeying 'the decrees of pontifical law', are made to seem remarkable, promising as they do a swifter 'gift of final peace'.[53]

Already by this stage the moral of the play is clear. We have a choice: we can either live up to the promise and accept the gift, or we can opt to go the way of madness. In Act Four (where, besides the Orkneys, Giovio also briefly talks about the Hebrides, the Shetlands, Thule and part of Norway) the choice is then spelt out. On the one hand it can lead us to the Shetlands, or else it can lead us to Norway. In the Shetlands, for all their darkness and their fog (which remind us of the voyage of St Brendan), we are made to catch a glimpse of the 'felicity' that is the lot of those who dwell among the blessed on some fortunate isles: their inhabitants, we are told, 'enjoy justice and peace and, […] since they have never even heard of the name of wealth or lust, they live in extreme poverty and in fog's perpetual darkness, with almost all of them, by an incredible bounty of nature, reaching a ripe old age'. In Norway, on the other hand, we are reminded that otherwise our destiny could be the loss of our humanity, for there we see 'atrocious Scythian beasts' who, 'covered all over in the longest of hairs', feed on travellers 'in the fashion of wild cannibals'. But should we fail to make the choice, we will for sure succumb anew to the demands of history which, as the work concludes, once again raises its head, threatening to take away for good our pleasure and our *otium* and our fables:

> but to writers telling of such monstrous places and such uncommon dangers nobody, unless they be shameless, would give any credence or would dishonestly afford any space. We, for our part, as we rely on the shining example of recent history, itself a dim reflection of the truth, are not to be seen to be writing simply for the sake of pleasure, offering sweet nothings to the sick or fables to the idle [*ociosi*].[54]

The dream, it seems, was pointless after all. There were places which even cosmography could not reach and speaking metaphorically, all things considered, was no alternative to speaking historically.

And yet if history offers but a 'dim reflection of the truth', what is the point of speaking historically? Will those *for* whom it is intended be satisfied and 'gift palaces and noble mansions' to 'the honoured hands of writers'? Or before they do so, will they insist on knowing where precisely the truth lies? The conundrum is truly Ariostesque. But what it teaches us is that upturning the truth, and deconstructing what has already been constructed by others, is not as simple or as effective as St John had made it out to be. Neither is it any easier to attempt to align the truth of the past with the truth of the present, and vice versa, or to separate chorography from chronology. Doing so produces Irishmen who sit uncomfortably astride the fault-lines of a divided world: 'hairy beasts' who, by virtue of an oxymoron, are forever destined to succumb to the demands of 'empire' and 'justice', and Irish 'kinglets' who, by virtue of an *adunaton*, might help the 'arms of Rome' to 'stretch to the horizon'. But perhaps that is what made *Ibernia* truly *fabulosa*. By being both real and unreal, it could be all things to all men (and some women). It could be both of this world and of the other, upside down and downside up, and in the end the truth of it, as we have suggested repeatedly, depended on the want of those *for* whom it was constructed. In any case what was the difference between a history which left Henry VIII undead, as did Virgili's, and a chorography which did the same?

Notes

1. 'E se tu vuoi che 'l ver non ti sia ascoso,/Tutta al contrario l'istoria converti;/Che i Greci rotti, e che Troia vittrice,/E che Penelopea fu meretrice.//[...] al vostro mondo fui scrittore anch'io' (ARIOSTO (1959), p. 378: 35, 27–8).
2. 'Non sí pietoso Enea, né forte Achille/Fu come è fama, né sí fiero Ettorre;/E ne son stati, e mille e mille e mille/Che lor si puon con verità anteporre:/Ma i donati palazzi e le gran ville/Dai descendenti lor, gli ha fatto porre/In questi senza fin sublimi onori/Da l'onorate man degli scrittori' (ARIOSTO (1959), p. 378: 35, 25).
3. The story of Ruggiero's rescue of Angelica is told in *Orlando furioso* 10, 91–115 (ARIOSTO (1959), pp. 91–3); 'di eterna gloria degno' (ibid., p. 87: 10, 57); 'Italia torneran nei primi onori' (ibid., p. 59: 7, 61).
4. 'Vana religion' – 'Eran senza contrasto quei d'Irlanda/Da piú parte ne l'isola saliti:/E spenta ogni pietà, strage nefanda/Di quel popul facean per tutti i liti;/Fosse iustizia o fosse crudeltade,/Né sesso riguardavano né etade.//[...] L'aver fu messo a sacco; messo fuoco/Fu ne le case; il populo fu ucciso;/Le mura fur tutte adeguate al suolo,/Non fu lasciato vivo un capo solo' (ARIOSTO (1959), pp. 98–9: 11, 46; 52–3).
5. 'Gran regina' – 'A pena un giorno si fermò in Irlanda;/Non valser preghi a far che piú vi stesse' (ARIOSTO (1959), pp. 101–2: 11, 80; 78).
6. On the additions to the third edition of *Orlando Furioso* see Casadei (1988).
7. 'Quando Eva ingannò col melo' – 'Come trovasti, o scelerata e brutta/Invenzion, mai loco in uman core?/Per te la militar gloria è distrutta,/Per te il mestier de l'arme è senza onore;/Per te è il valore e la virtù ridutta,/Che spesso par del buono il rio migliore' (ARIOSTO (1959), p. 96: 11, 22; 26).
8. On Virgili see Bacchielli (2003), Hay (1952: 'justifying', p. 9). The full Latin text, with English translation, of Virgili's description of Ireland, as it appears in the manuscript as well as the printed versions, can be found in VIRGILI (2009)/Haywood (2009b). Since it is easily accessible on-line, I will not here provide the Latin original of the quotes but simply refer to the relevant page (and section) numbers of the article.
9. On Giovio see Agosti (2008), Harris (2007), Haywood (1998), Michelacci (2004), Mayer (1985), and especially Zimmermann (1995: key in well of St Patrick, p. 174).
10. 'Er konnte über alles schreiben; aber er verstand nichts' (Fueter (1911), p. 54).
11. VIRGILI (2009): pp. 220–1 (1), 228–9 (14), 230–1 (16). There are modern editions of Giovio's description of Ireland in GIOVIO (1984) and Harris (2007),

both of which contain some errors (as does Harris's translation). References are to the edition in GIOVIO (1984), corrected on the basis of the 1548 original where necessary, and with punctuation occasionally altered.

12. Writing as an Italian: see Cespedes (1979), p. 380. John Bale: see Hay (1952), p. 158. 'Anche l'Inghilterra dovette la prima compita Storia che di quell'isola venisse a luce, a un Italiano [...]. Ma [...] ch'ei gittasse al fuoco le antiche Cronache, delle quali si era giovato, perché perdendosene la memoria, la sola sua Storia corresse per le mani de' dotti, non vi sarà uom di buon senso, che nol creda una favola' (Tiraboschi (1784), pp. 369–70). Exercise of logic, etc.: see Levy (1967), pp. 58–9, 63–4, 293.
13. VIRGILI (2009): pp. 224–7 (12, 13). GIOVIO (1984): 'Nec dimidiam quidem Britanniae partem aequare existimatur [...] et, quod Angli facile concedunt, tepore soli, coeli clementia atque aeris serenitate nobilior' (p. 119). On the shape of Italy (oak-leaf or boot?) see Defilippis (1993), p. 28.
14. VIRGILI (2009): pp. 226–7 (14). GIOVIO (1984): 'Hybernia certe miraculo quodam serpentibus caret, neque ullum noxium animal ea terra patitur praeter lupum; caetera uti apud nos cernuntur praeter asinos, quos vel importatos haud dubie non alit' (p. 121).
15. GIOVIO (1984): 'Equi autem tota Hybernia tollutarii incorrupta sobole progignuntur, quos a mollissimo incessu hobinos Angli vocant; ob idque delicatis expetuntur et in Gallia Italiaque nobilioribus foeminis dono dantur. Ex hoc genere duodecim, candoris eximii, purpura et argenteis phaleris exornatos, in pompa summorum pontificum sessore vacuos duci videmus; iusta proceritas illis deest, sed forma, decore, spiritu maxime praecellunt. Holtonia autem peracreis, robustos et vigore animi prorsus intrepidos, tanquam ad bella natos, producit, quibus post ordei pabulique mensuram acrifolii frondes in flamma perustas, et crepitantes in praesepibus apponunt [*recte* opponunt]; quo alimento sanari convulsos et vitalia foveri experimento didicerunt. Hos Hispaniae Galliaeque mercatoribus permutatione aut pecunia divendunt' (pp. 121–2).
16. VIRGILI (2009): pp. 226–7 (14). GIOVIO (1984): 'margaritas, pondere levoreque, si candor accederet, Indicis pares' (p. 121).
17. GIOVIO (1984): 'abundat salmone pisce aringisque plurimum, quibus sale et fumo inveteratis mercimonii nomine Italiam repleri conspicimus' – 'Obocus, qui hodie Lipheus dicitur, sinuato cursu [...] Dublinium interfluit atque inde [...] magnarum navium capax [...]' – 'insignis Suirus, [...] auctus [...] duobus in unum confluentibus fluviis, Vaterfordiae urbis portum tutissimae stationis Anglorum et Belgarum classibus praebet' – '... martarum et sylvestrium fellium pelles, quibus nobilium fulcitur vestis, taurorum item tergora et lanatas ovium exuvias et qui a volatus pernicitate maxime probantur falcones regum aucupiis

dicatos; pro his ab Hispania Galliaque loricas, galeas, spicula, gladios referunt, atque item vinum, oleum et crocum' (pp. 121–2).

18. On the miracles of Ireland Virgili (2009), p. 234 (25) writes: 'Miracula autem Hyberniae, quae vulgo praedicantur, quoniam eo remotius fugere solent, quo diligentius inquiruntur [MS: investigentur?], repetere supervacaneum duximus'. Regarding the barnacle geese (which he had looked for in Scotland), Pius had written: 'de qua re cum auidius inuestigaremus, didicimus miracula semper remotius fugere' (Pius (2001), p. 186).
19. Virgili (2009): pp. 232–3 (22–3).
20. Giovio (1984): 'Appellantibus festive arrident et non numquam choreas ad arpen ducunt, quam utroque genu contentam per aereas impares chordas argutis digitis pulsant, et modulos alludente voce comitantur' (p. 120).
21. Giovio (1984): 'stagnum haud peramplum admirabilis naturae celebratur' – 'miraculo [...] proximum videtur' – 'De oraculo autem ad divi Patritii puteum, quo novenis haustibus expiari animos a gravibus delictis ferunt atque inde concepto vaticinio ad divinationem praeparari, nihil hactenus a nobis compertum est, nisi cuniculum illum in abdita rupe ad extremam insulae oram, qua cori flatibus exponitur, antiquorum operis excavatum, ubi puri laticis vena scaturiat, inter miracula reponamus. Nos haec uti fabulosa curiosae superstitioni et peregrinantium votis relinquemus' (p. 121).
22. Giovio (1984): 'in multas regiones [...] maxime [...] insignes' – 'sacrae quattuor metropoles' – 'Dublinium [...] amplissima omnium civitas [...] Anglorum enim commerciis maxime crevit' (pp. 120–1).
23. Virgili (2009): pp. 234–5 (24, 27), 226–7 (15).
24. Giovio (1984): 'Regia Honelis Armacana dicitur, quae praecipua sacrorum authoritate totius Hyberniae primatum obtinet; sed nullis cingitur moenibus, sicuti et caetera Holtoniae castella nudata muris patent, quum ditionem adversus hostes expedita manu campestrique acie defendere quam munitione atque praesidiis tueri nobilius ducat. Vicatim habitant et in deligendis sedibus opportunitates paludum et montium sequuntur. Connatius, uti bellicosum ac intentum decet, pagos devitans, in stativis [*recte?* aestivis] castris ac hybernis vitam traducit. Umbraculis fronde protectis tentoriisque ex pellibus et lintris [*recte* linteis] militari more imbres arcet' (p. 120).
25. Virgili (2009): pp. 220–1 (1–2).
26. Virgili (2009): pp. 220–1 (3), 230–1 (16).
27. Virgili (2009): pp. 232–5 (22, 24).
28. Giovio (1984): 'Interemptorum funera non planctu et lachrymis sed decantatis laudibus celebrantur; praeficas enim propinqui precio conducunt, quae exequias naeniis exornent. Pertinere siquidem ad familiae decus arbitrantur,

egregie fato functos non diu luxisse, quod non lachrymis sed diuturna memoria dilectos prosequi praestantius ducant, tanquam eis honestum vitae exitum illudque incomparabile bonum supremae pacis invidisse dispudeat' (p. 122).

29. GIOVIO (1984): 'Hyberni omnes vel in asperis et sylvestribus locis christianos observant ritus. Docentur ab episcopis per idoneos literarum latinarum et musicae artis magistros, qua institutione eruditioneque opima sacerdotia promerentur' – 'Lites non diu alere mos est, quum advocatis non sit locus et controversiae aut expedito iudicio aut armis finiantur. Soli sacerdotes antistitum tribunalia frequentant, utpote qui pontificii iuris decretis pareant' (pp. 121–2).

30. VIRGILI (2009): pp. 220–1 (3). GIOVIO (1984): 'Apud eas in universum dubius est pudor et rara admodum pudicitiae laus, quum in peramplis cubilibus integrae familiae cubent et facile tenebris verecundiam extinguant. Reguli nobilesque uxorem pariter et pellices alunt, neque probro ducitur uxorem repudiasse, redditoque ei mundo muliebri, eundem nuptialem annulum concubinae digitis induisse. Porro alii veneri licentius indulgent, neque libidinem inter gravia crimina christianis vetita legibus adnumerant, quum probitatis famam una praesertim innocentia metiantur' (p. 120).

31. GIOVIO (1984): '... lanea chlamys [...] limbo ad elegantiam variegato iubatoque. In ea princeps puniceum colorem praefert, militi cianeum et ruffulum relinquit. Iccirco nostri Hybernicam appellant, uti inde traductam, quam trierarchis et navalibus sociis familiarem videmus. Tunicis etiam croceis usque ad genua conteguntur, quarum margines inductis serici coloribus acu scitissime depingunt' (p. 120). VIRGILI (2009): pp. 230–1 (17–18).

32. GIOVIO (1984): 'Puellis mos est crines nodulis collectos ostentare, velare autem matronis, quae croceis fasciis per multas spiras convolutis in orbem capita Turcarum more exornant' – 'Horrido enim et circumtonso ad aures capillo, recisaque saepe barba, et labro superiore intortis setis ad terrorem hirsuto populariter incedunt, ut bello ferociores appareant' (p. 120). VIRGILI (2009): pp. 230–1 (19).

33. VIRGILI (2009): pp. 232–3 (23). GIOVIO (1984): 'Verum inculta gens, ignara luxus et peregrinis illecebris incorrupta, aratrum magnopere fastidit, ligonem recusat sudoremque omnem praeter bellicum refugit, quo uno nihil fere, praeter decus, quaerendum existimat [...]. Quin, et eodem contumacis animi decreto metalli feris montibus parcunt, nec aurum argentumve secturis et fodinis Anglorum studio et multo labore inde erutum asportari patiuntur, uti rem vitae ac moribus maxime pestilentem' (p. 119).

34. GIOVIO (1984): 'Contenti siquidem pecore, lacte, caseo et favis, pisce item, aucupio venationeque, dona Cereris et Bacchi principum suorum illata mensis non invident, quum vulgo placentas ex olira et birram in poculis praeferant, et multo legumine largiter epulentur' (p. 119). Ceres: Roman goddess of agriculture (cereals).

Bacchus: Roman god of wine. VIRGILI (2009), pp. 228–9 (15). In the manuscript he had written: 'ut durissimum hominum genus carent lacte et pomis victitant, panem siquando vescuntur, ex farina avenacea conficiunt'. For the printed edition he wrote: 'ut longe durissimum hominum genus non multum frumento, sed maximam partem lacte atque pecore vivunt, panem quando vescuntur, ex farina avenacea conficiunt': pp. 228–9 (15). CAESAR (1970): 'Neque multum frumento, sed maximam partem lacte atque pecore vivunt' (p. 180: 4, 1).

35. 'Cum praesertim tu rectum omnino gestarum rerum aeque ut gerendarum scias iudicium facere, quippe qui, ad imperium & iustitiam natus, magna praeditus es prudentia literisque pariter divinis atque humanis apprime eruditus cunctas bonas calles artes. Ac ita tua singulari virtute non modo eorum Principum qui hodie sunt gloriam, verumetiam omnis antiquitatis memoriam facile superas' (VIRGILI (2009), p. 2).
36. VIRGILI (2009): pp. 234–5 (26).
37. VIRGILI (2009): pp. 228–9 (14: emphasis added), 236–7 (28).
38. VIRGILI (2009): pp. 232–3 (21).
39. VIRGILI (2009): pp. 236–7 (30: emphasis added).
40. GIOVIO (1984): 'fuit aliquando tempus quo Angli, magnis exercitibus insulam pervagati, ad totius Hyberniae principatum aspirarent, sed Hyberni, qui diu intestino occupati bello graves insaniae poenas dederant, quum periculo admoniti ad publicae salutis consilium rediissent, coniunctis viribus, ita Anglorum opes atque animos contuderunt, ut ne illis exigua pars insulae, et ea quidem liberali quadam pactione, relinqueretur. Armavit enim Hybernia Anglico tumultu supra quadraginta millia peditum et equitum' (p. 120).
41. GIOVIO (1984): 'Hyberni […] ex aequo suis regulis parent, neque unquam passi sunt, ut summa imperii ad unius arbitrium, veluti sublata libertate, recideret. Usqueadeo in feris hominibus ingenium viget, ut iustitiae nomine plurimum gaudeant, idque unum maxime curent, ne quis multo maximis viribus abundet indeque, uti plerunque fit, intolerandus et impotens, caeteris subactis, insulae regnum arripiat' (p. 119).
42. GIOVIO (1984): 'Verum ingenti authoritate, bellicis operibus et antiqua pariter nobilitate caeteros antecellit Connatius Honel, Holtoniae princeps. Is continenter bella gerit, modo cum finitimis Anglis, modo cum regulis, qui undique Holtanicos fines amplectuntur' (p. 119). On Conn O'Neill see Ellis (1985), *passim*, Harris (2007), pp. 272–5, Hayes-McCoy (1976), *passim*, Moody Martin and Byrne (1976), pp. 44–52, 66–8, 71–3.
43. GIOVIO (1984): 'In summo igitur honore sunt milites, equitesque in primis et secundum hos gravis armaturae pedites, quos, uti veteranos et bella ex bellis ferentes [*recte* serentes] in castrisque educatos, ipsorum lingua Galloglathas vocant.

Inter omnes Connatius Honel equitatu pollet et gravis armaturae pedites unus alit; cum iis enim agros et fortunas omnes partitur, quum pecuniae egens ipsis vitae alimentis ac item veste, armis et equis stipendia persolvat. Eques lorica et galea munitur; laeva lanceam Hispanicam habenasque simul comprehendit, dextra validissime amentata iacula contorquet; plura enim aut contenta gladii baltheo aut compressa foemore gestare consuevit, alia pedites famuli subministrant. Ita eminus iaculis rem gerit, cominus autem, traducta celeriter e leva in dextram lancea atque ea ad medium alte sublata, iteratis ictibus pugnat. Binos singuli trahunt equos, alterum quo extra aciem vehuntur, alterum autem egregium bellatorem sessore vacuum, in quem opportune transcendit, quum classicum ad arma vocat. Non clangore tubae sed ingentis ligneae buccinae mugitu praelium accendunt. Tanta vero equiti ad regendum atque incitandum equum agilitas, ut nutabundus et eludens miro corporis flexu hostilia tela devitet atque ea solo sparsa poplite ab ephippio suspensus exporrecta laeva corripiat. Veterano autem pediti praeter galeam et loricam lata ex maculis ferreis fascia guttur munit; is tum Germanica secure utitur, cum initio praelii iaculum emiserit. Ex his stataria acie in medio constituta, praecinente utriculario tibicine, levis armatura e cornibus procurrit, iaculis sagittisque res geritur, quas brevibus et ligneis arcubus emittunt. Galloglatha autem tanta constantia depugnat, ut nullo mortis metu fortiter vincendum aut in vestigio honeste cadendum putet' (p. 122).

44. Giovio (1984): 'Duae siquidem maritimae regiones in Laginia (Midiam et Fingalliam eas vocant) Angliae regum imperio et praesidiis tenentur. Ab his brevissimus est in Angliam traiectus, intercedente Mona insula, in quam, authore Tacito, repertis humilibus vadis captatoque aestu reciprocantis Oceani, miraculo quodam Romanae legiones transnatarunt (ea vetustum adhuc immutata littera nomen retinet), ita ut a Mona Hyberniam invadere in spem triumphi ambitiose cogitarint, putaritque Iulius Agricola legione una et modicis auxiliis subigi ac in pace contineri facile posse' (pp. 119–20).
45. Giovio (1984): 'Ipse nuper Connatius Honel ex provincialibus suis, exquisito etiam delectu, quattuor millia equitum et triplo maiorem peditum numerum in aciem eduxit; sed demum Thomae Havardi, Scotica victoria clarissimi, industria et donis in amicitiam traductus, cum Henrico rege pacem colit' (p. 120). Thomas Howard was the Duke of Norfolk: for a discussion of this see Harris (2007), pp. 273–5.
46. Virgili (2009): pp. 234–5 (25).
47. Cicero: 'Rerum ratio ordinem temporum desiderat, regionum descriptionem' (Cicero (1967), pp. 244–5: 2, 15, 63–4). Pontano (1943a), pp. 123–239. Virgili: 'Ipsa autem historiae aedificatio posita est in rebus et verbis. Rerum enim ratio tam temporis ordinem quam locorum descriptionem poscit' (Virgili (2002),

pp. 110–13: where 'locorum descriptio' ['description of places'] is translated as 'geography'. 'Chorography' would be a more accurate translation).

48. 'Describit Africae situm Crispus, quo scilicet et situs ipse notus esset et qui amici quive inimici ex omni provincia' (PONTANO (1943a), p. 219). PIUS (1571?), p. 282: 'rerum quas scribimus [...] campus est ipse terrarum orbis'.

49. GIOVIO (1984): 'Europam itaque, antiquorum sequentes ordinem, primo describemus, incohantes a Britannia, quanquam iure optimo ab Italia, quondam gentium omnium victrice, ipsaque ab aeterno fatalique imperio augusta urbe initium capere deberemus' – 'Itaque, sepositis historiis, atque animo ad hilaritatem composito, a via veterum studiorum non inepte deflectemus, vertemurque molliore stilo ad iucundissimum scribendi genus, ut non studiosis modo, sed delicatis atque ocio languentibus, peramoeni pulcherrimique operis novitate voluptatem afferamus' (pp. 89–90).

50. In a letter to Pier Luigi Farnese, Duke of Piacenza, dated 21 February 1547, discussing what would happen following the death of Henry VIII, Giovio writes: 'What Our Lord [Pope Paul III] should do now I could not tell, except that it would be very appropriate for the Holy See to seize the opportunity of sending Reginald Pole, of the House of York, connected to that of Lancaster, to take that darned virgin [Princess Mary], defrocking him and sending him with the help of France and the help of the Emperor' ['Quel mò che avesse a fare Nostro Signore non saprei dire, se non che sarebbe molto a proposito alla Sede apostolica il prender l'occasione di mandar Rainaldo Polo della Casa degli Orchi, congiunta a quella de Lincastro, a prendere quella verginaccia, disdiaconandolo, e mandandolo con agiuto di Francia e con agiuto del Imperatore': GIOVIO (1956), p. 70].

51. GIOVIO (1984): 'Nihil autem ea iuventutis educatione institutioneque modestius religiosius atque frugalius, ut non mirum sit inde provenisse ingenia quae in Dialectica, Philosophiaque et in sacris demum literis Europam omnem subtilitatis admiratione compleverint, et tum maxime quum maiores nostri, repudiatis optimarum scientiarum antiquis authoribus, nihil potius quam difficiles captionum nodos, et convoluta disputantium sophismata mirarentur' – 'luxu et perpetuis gulae voluptatibus dediti' – '... lex iudicio maxime iniquo atque impio conscripta [...] supra religionis causam [...] antiqua etiam foedera [...] execrabili temeritate turbabantur' – 'tam scelesto quam ambitioso consilio' – 'Quanquam enim tot nuptias, quibus nullum adhuc finem videmus, impotenti libidine repetitas, clarissimorum ac innocentissimorum virorum sanguine defoedarit, omnia tamen ei foeliciter eventura crediderim, si saluberrimum volumen a Reginaldo Polo cardinale propinquo suo, ad se pie amantissimeque perscriptum, pari pietate aequitateque animi opportunae medicinae instar hauserit' (pp. 99, 101, 107, 109–10).

52. Giovio (1984): 'pervetusta [...] consuetudine [...] tumultu hostili' – 'post multa ex intestinis caedibus interregna, atrocesque multorum tyrannorum interitus' – 'ambitio regnique cupiditas' – 'epularum autem copia edacitateque nostris hominibus pene incredibili Anglos antecedunt' (pp. 111–14).
53. Giovio (1984): 'scelesta rapacita[s]' – '... ut mira ingenii foecunditate librum adversus Lutherum heresiarcham publicaret' – 'Fuit ea lex iudicio maxime iniquo atque impio conscripta promulgataque [...] antiqua etiam foedera, inviolata ante tot regibus, execrabili temeritate turbabantur' (pp. 104, 107).
54. Giovio (1984): 'Iustitia et pace gaudent, et quum divitiarum atque luxuriae nomina numquam audierint, in summaque egestate et in perpetuis brumae tenebris vivant, fere omnes incredibili naturae foelicitate ad summam senectutem perveniunt' – 'Scythicas atroces belluas' – 'praelongis undique setis et crinibus coopertos' – 'immanium Canibalium more' – 'Sed haec monstrifera loca inusitataque pericula narrantibus, nemo nisi inverecundus fidem astruxerit, aut improbus elevarit. Nos autem in praeclaro testimonio recentis historiae, uti incertissima veritatis luce consistemus, nec adulatione dulcissima aegris aut ociosis animis fabulosa ad voluptatem scribere videamur' (pp. 124–5).

CHAPTER SIX

New World
Imagining Ireland in a New World

Authors and works discussed in this chapter (in chronological order):

- Benedetto Bordone or Bordon (1450–1530), *The Book of Benedetto Bordone in which Are Discussed All the Islands of the World* [*Libro di Benedetto Bordone nel quale si ragiona di tutte le isole del mondo*]
- Pietro Martire d'Anghiera, aka Peter Martyr d'Anghiera (1457–1526), *Summary of the General History of the West Indies Extracted from the Books Written by Master Peter Martyr* [*Summario de la generale historia dell'Indie occidentali cavato dai libri scritti dal signor Pietro Martyre*]
- Sebastian Münster (1488–1552), *Cosmography or Description of Every Land* [*Cosmographey oder Beschreibung aller Länder*]
- Tommaso Porcacchi (1530–85), *The Most Famous Islands of the World* [*L'isole più famose del mondo*]
- Fra Egnazio Danti (1537–86), Map of Ireland, Florence, Palazzo Vecchio, Guardaroba Nuova
- Giovanni Lorenzo d'Anania (c.1544–c.1608), *The Entire Edifice of the World* [*La universal fabrica del mondo*]
- Giovanni Botero (1544–1617), *Universal Reports* [*Relationi universali*]

Giovio's attempt to separate cosmography from history may not have worked out quite as he had hoped, but his instinct had been right. Cosmography was coming into its own and the demand was now growing for works which focused on the description of lands and their inhabitants with only minimal reference to historical events, or with history so potted that it lost any semblance of the truth and 'good-forness' it was meant to deliver. Seemingly what mattered most were now the bare facts, and with the world having changed beyond recognition – certainly beyond the recognition of those who still placed their faith in ancient Romans and Greeks (Africa and Asia disjoined for good, the Indies of the west emerging as a fourth continent, etc.) – and geography, thanks to the advances in cartography and to the testimonies of an ever-increasing number of eye-witnesses, having acquired the potential to become ever more Ptolemized (mimetic) and ever less Strabonified (*moralisé*), the balance could now shift and the paradox mentioned at the beginning of Chapter Two be resolved in favour of truth *about* instead of truth *for*. Yet what is surprising – or perhaps it is not – is just how imperious the latter continued to be. There was so much more to know in the world, but the need not to know, or to know only what mattered most, was stronger than ever. Facts that were as bare as Chiericati's happy natives were just too unsettling to contemplate, it would seem.

A telling example of this can be found in the *Summary of the General History of the West Indies Extracted from the Books Written by Master Peter Martyr* [*Summario de la generale historia dell'Indie occidentali cavato dai libri scritti dal signor Pietro Martyre*] published in Venice in 1534, in the same year, that is, as the first edition of Virgili's *Anglica historia*. Peter Martyr (Pietro Martire d'Anghiera: 1457–1526) was an Italian who entered the service of the kings of Spain, rising under Charles V to become royal chronicler and a member of the body responsible for the government of the Spanish colonies in America, and he is principally remembered today as the author of a massive work *On the New World* [*De orbo novo*], published by instalments, in Italy, Spain and Basel, between 1504 and 1530. The *Summary* is an abridgement of the work, intended for Italian readers too busy to read it in its entirety, and possibly too limited to grasp it fully. Towards the end of the abridgement, there is a description of the inhabitants of the furthest province of the island of Hispaniola. The people who live there, Martire had written, are so savage that they have never wanted to have contact with the other inhabitants of the island. Such is their savagery that they live in caverns, in woods and on 'very high mountains', and only feed on wild fruit; and what is more, they do not have a 'specific way of talking among themselves as have all the other men in the world'. Neither do they know what 'master' means, and they have no laws, 'but are totally wild animals, except that they have a human appearance'. Sometimes they can even be seen wandering about stark naked. And yet it is almost impossible to capture them, for they run faster even than the fastest dogs that have been imported into the island.[1]

One September, in a most beautiful valley, where some Christians had cleared many fields and had gathered with all their children, one of these wild men suddenly appeared out of the woods, 'tall and terrifying'. He grabbed a small child that was lying on the grass close to his father and ran away with it. The father and the other Christians all gave chase, shouting to high heaven, but the wild man, though he stopped and appeared to wait for them, soon disappeared – in order to devour the child, as far as the father was concerned. However the wild man, having spotted some shepherds who were grazing their flocks in a nearby valley, quietly walked towards them and left the child a short distance away, beside a road along which they were bound to pass, and that evening father and child were reunited. Nevertheless, and even though all was well that ended well, the editor of the abridgement clearly felt that his readers would find it impossible to comprehend such uncommon savagery. He therefore added the following explanation to the text:

> let the readers not be surprised that on this island, so far away from us, one finds this type of wild men, because on the island of Hibernia too, which is beneath the king of England, and not very far from there, in the midland part, where there is nothing else but woods and very high mountains, one knows that there is an infinity of wild men, who have never wanted to have intercourse with those who live by the sea, and whom it has never been possible for the said king's soldiers to defeat.[2]

In other words, the new world was not really new. All that had happened was that the borders of the old one had grown thicker and the fables of Hibernia taken on a wider resonance. The readers – *Pandineschi* readers at least – were not to be surprised.

What is for sure is that the readers cannot have been very discerning. They were not of the type, apparently, that liked to talk with books, preferring instead to let the books talk at them. Otherwise they would have noticed that those who wrote the books – books which often claimed explicitly to be telling them the whole truth and nothing but the truth – were not quite the experts they claimed to be. Their expertise, which, geographically speaking, was channelled into the production of three different genres (maps, island-books and cosmographies), had more to do with recycling, from the comfort of their armchairs, what was in the process, maybe, of becoming an *acquis* (at least with regard to Ireland) but which they did their very best to peddle as the emperor's new clothes. Clearly it was best to avert one's eyes from that type of bareness too. We today, however, cannot help but gawk and be surprised, or at least amused.

Of the three categories of self-styled experts mentioned above, mapmakers made forward strides like no other geographers in the course of the sixteenth century and undoubtedly came closest to delivering the scientific objectivity which we feel should be the hallmark of geography, or any other field of knowledge for that matter. Cartography, though, is a subject in its own right, which falls beyond the scope of this book. Here all that can be said is that talking maps of the kind designed by Fra Mauro, maps, that is, with explanatory captions, continued to be very popular, and one in particular deserves our attention. Between 1563 and 1565 the Duke of Florence and later Grand Duke of Tuscany, Cosimo I de' Medici (1519–37–74), prompted by his chief architect, Giorgio Vasari (1511–74), commissioned the Dominican mathematician and cosmographer Fra Egnazio Danti (1537–86), who was subsequently to oversee the execution for Pope Gregory XIII (1572–85) of the maps in the Vatican's Gallery of Maps, to cover the walls of a cabinet of curiosities in Florence's Palazzo Vecchio (the so-called Guardaroba Nuova) with a series of maps representing the different lands and islands of the world.

The maps are all accompanied by cartouches with brief descriptions of the lands in question and, arranged as they are in two superimposed rows, they cover the better part of all four walls. As a result some of them are less visible than others and some of the cartouches, in what is quite a dark room, could only have been read with the help of a ladder and a torch. The map of Ireland and of Britain, however, is clearly visible with the naked eye and occupies one of the most prominent positions on the wall, at shoulder level on the right-hand side of the door as you leave the room. And more visible than most is the cartouche concerning Ireland, which stands out in the top left-hand corner of the frame, where Britain's is tucked away in the bottom right. Whether this came about by command of Duke Cosimo, we do not know, but what is for sure is that every time he walked out of the cabinet, in the company of selected guests, the Duke would have found it hard not to catch a glimpse of the following:

> Hibernia [*Ibernia*], today called Ireland [*Irlanda*], is situated between England and Spain, and extends to about 280 miles in length and 100 in width. Together, its bishoprics and towns add up to 50, and its main rivers are the Suir [*Suiro*], the Boyne [*Boanda*] and the Shannon [*Sineo*], which divide it into four parts, namely Munster [*Momonia*], Leinster [*Laginia*], Ulster [*Hultonia*] and Connaught [*Connaccia*]. On this island the air is so temperate, that not only is no poisonous animal or plant born there, but if it is brought from elsewhere it does not adapt, but dries and dies. In the northern part of the island there is a very deep pit, which some serious writers say is the well of St Patrick, the bishop of the said island. Equally they claim that on one of the Orkney isles there is a tree, standing with its roots in the water and floating near the bank, whose fruits fall into the water and produce birds.[3]

Every time, that is, the Duke left the sanctuary of his inner world to face events in all their unpredictability, he would have been reminded that there was one thing at least which remained predictably the same, just as it had done for Fra Mauro, and that was Ireland's fabulousness. It now stood carved in stone, so to speak (in reality the maps are painted on wooden panels), and if 'some serious writers' said that it was so, then so it had to be.

The immutability of Ireland is all the more remarkable in that the cartouche concerning Britain – to the west of which lies '*Hibernia*, today called *Irlanda*, and by the English *Irlond*' – appears to have been deliberately designed to draw attention to the contingency of our knowledge of the world, noting as it does that 'Caesar in his commentaries says that the said island is 2,000 miles in circumference and the Venerable Bede 2,600 but the moderns (who are to be believed more, following the teaching of Ptolemy) claim it is no more than 1,700 miles'. The cartouche also notes that the inhabitants of Britain were considered by the ancients not to have any 'letters or music', but that 'today' they are seen to be 'most excellent' at both, besides being 'valorous in arms' and much concerned with 'nobility' and 'splendour'; however 'now they are all heretics and separated from the Catholic Church'.[4] In other words, no more than Virgili had been able to, did Danti manage to keep the 'now' at bay; the demands of God and Caesar would not allow him to. Ireland, though, was in a different category. With Conns and Obertos having served their purpose, it would seem, its chorography could once again be kept apart from its chronology. That bit of the wall, at least, would never need to be painted over.

A similar process was at work in island-books – or so it would appear. When Silvestri had put together his, torn between the 'amazing to say' and the 'worth remembering', he was racked by doubts, not least regarding Ireland, with its suggestion that 'all things change with time'. Still he tried as best he could to make his Ireland season-proof, by not asking Bishop Milo too many questions that might deflate his wonderment. A century and a half later, when the first commercial island-book saw the light of day, the doubts and wonder and the need to question have been set aside. Its author, it is true, is well aware that time does not stand still, but the awareness in his case is no more than an alibi for self-promotion. Berating the ancients for having failed to 'investigate any further than that which they found to be written by those ancient writers of theirs' and thus for not having managed to discover 'the islands which were found in our own days', he proudly congratulates the 'men of our own times' who, 'notwithstanding the very great loss of wealth, or the loss of their own lives, did not fail to explore the world in all its parts, investigating the whole circuit of the earth', and confidently proclaims that, as a result, he is now in a position to correct 'the many errors which the ancients had left to posterity'. In order 'to teach those who do not know' and to provide pleasure to those 'who are occupied in other studies, and those too who are not inclined to sailing', he therefore promises to recount everything there is to know about islands. His book, he declares, will describe

> all the islands and peninsulas of the world, with their ancient names and modern ones, and with every other thing that pertains to them, the stories that are written about them, as also their fables, and in what part of the sea they lie, and of the varying customs which, sailing all the while, are found in them, and beneath what parallel and in what climate they are positioned.[5]

The readers, it is clear, were not to feel short-changed.

Perhaps they might well have done had they known that the author of these words was not someone who had 'sailed all the while'. He too was an armchair traveller. His name is Benedetto Bordone or Bordon (1450–1530) and the furthest he ever ventured from his native Padua was to Venice, twenty miles up the road. His father was a barber, his grandfather a tailor and his son, who tried as best he could to conceal his humble origins, the famous polymath Giulio Cesare Scaligero or della Scala (1484–1558), but apart from that not much is known about him, except that he was self-taught and that he mostly lived by his quill, in the orbit of Venice's flourishing book industry, gaining quite a reputation as an illuminator of books and manuscripts. Coincidentally, one of the more famous manuscripts he decorated is now housed in Dublin's Chester Beatty Library. His real passion, however, was geography, and like Berlinghieri he spent the better part of his life acquiring the knowledge he would later distil in his major work. Unlike Berlinghieri, though, he had a good understanding of the needs of the reading public, and after its first edition, in 1528, *The Book of Benedetto Bordone in which Are Discussed All the Islands of the World* [*Libro di Benedetto Bordone nel quale si ragiona di tutte le isole del mondo*] was republished another five times (1534, 1538, 1540, 1547, 1565), and for close on half a century it remained the standard reference work on the matter.[6]

It is not a very long book. It is made up of three sections – one for the islands of the Atlantic, including those of the New World, one for the islands of the Mediterranean, and one for the islands of 'the Oriental ocean and Indian sea' – each consisting of a series of short, self-contained chapters, all of which deal with a separate island. Every chapter is accompanied by a map, which is often no more than a sketch, and the book also includes a number of world and regional maps, which on the whole are fairly well executed, following the principles of Ptolemy. The spirit of Ptolemy, together with that of Strabo, is also invoked, albeit anonymously, at the beginning of the work, where Bordone justifies the order of precedence he adopted. 'Since I must give a start to this discussion of the islands of the world', he announces there,

> it seems to me a seemly thing to begin from one of the ends of our continent, even though Europe is smaller than the other two parts; yet in strength, thought and wisdom it has always exceeded the other two parts, and therefore from this part, as from the noblest part of the world, I will make my start.[7]

But though the ancients stand as guarantors of Europe's right to matter most, in Bordone's eyes they also stand accused of having committed 'many errors'. He is thus in need of more credible patrons to vouchsafe for him and conceal his own limitations, and so he decides to place his work under the protective wing of a modern who could claim to be an 'outstanding connoisseur' in the matter of islands, and in addition could stand proxy for the mighty rulers of the waves. That person is his nephew, the 'excellent surgeon' Messer Baldassar Bordone, who has 'sailed the entire Mediterranean sea upon the mighty galleys of the Venetian lords and the Catholic King'. Yet even that is not quite good enough and, like Silvestri and Fra Mauro, Bordone also feels the need to enlist the support of an even mightier authority. Since what takes place on earth, he argues, is determined by the movement of the heavens, cosmography is a branch of astronomy. From astronomy we know that the earth is the measure of all things and that the heavens move in harmony. And harmony, we understand, is the fruit of God's own wisdom, thanks to which the order of the universe is unchanging. 'Consequently, since astronomy depends on the heavens, it is a perfect thing, therefore cosmography, since it depends on astronomy, is a true thing'.[8]

Nevertheless, though Bordone had God on his side, he clearly did not feel as confident as he claimed to be. In his chapter on Scotland, following Pope Pius's *De Europa*, he refers to the trees

> that can be found on the banks of a river and which produce fruits similar to ducks, which, when they are almost ripe, fall by themselves, some into the water, some onto the ground, and those that fall onto the ground rot and putrefy, but those that fall into the water become feathered birds which, when they have grown, fly through the air like other birds.

Going one better than Pius, however, he claims that it has been 'ascertained' that 'this miracle was to be found not in Scotland but on the Orkney isles' – where Pius had simply said that 'we learnt that miracles always flee further away and that the famous tree was to be found not in Scotland but on the Orkney isles'.[9] Still, ascertain though he may, Bordone is all too aware that his reputation may be at stake, and so, returning to the matter in the conclusion to the work, he seeks to cover up his vulnerability by exposing that of others. This miracle, he tells his readers, is as well as written black on white: and who would dare deny what some have seen with their own eyes?

> Who does not know that it would not be easy to make anyone who had not seen an Ethiopian believe that a man can be black? But often nature produces things which seem impossible, and yet they are. Something which is even less easy to believe is that the leaves of some trees, falling into the water, become feathered birds; such birds however can be seen in Venice, at Messer Andrea Rossi's, who had them brought over from Spain, and they are smaller than geese but larger than ducks, and the Spanish call them *gravagne*.[10]

Who Messer Andrea Rossi is, we do not know, but at least we can take it for a fact that Ireland's wild geese, having fled to Scotland and the Orkneys, were later spotted in Spain and even in Venice. If Bordone said so, then so it had to be. Cosmography, after all, was a 'true thing', especially if it followed in the wake of the rulers of the waves.

It also helped, of course, if cosmography could avail of the backing of men of learning, for all that they were prone to committing 'many errors', and what Bordone also said – following Pius and Strabo – was that there was nothing in Ireland

> which is worthy of being remembered, except that these islanders are very great eaters, and in their food they make great use of human flesh, and when their relatives are dead, they eat them, and this to them is a very great honour. Nor do they appreciate any less to mix in public with their women, both their mothers and their sisters.[11]

However, having staked his claim to fame on the correction of those 'many errors', Bordone could manifestly not leave it at that. He was duty-bound to find more in Ireland that was worth remembering, at the very least by supplementing the diet of the Irish with less recycled fare, although he could not, of course, allow the outer reaches of the world's 'noblest part' to be confused with the noblest of the noblest. The diet of the Irish, in whom 'there is more of the savage than there is in the British', is therefore made more varied, but in a way which is pointedly non-Mediterranean. The island, we are told, 'does not produce oil, wine or grain', but its inhabitants 'drink beer and use bread made from the flour of barley and rye'. Nevertheless, what they themselves did not produce, the Irish could possibly import, for Bordone does admit – which is no mean compliment, coming from a Venetian – that 'they are good merchants'. There is not much else, however, which he finds in their favour. The island is 'flat', he claims, and has 'an abundance of rivers', but, 'according to what Pius says', all the rivers are 'small'.

Pius, as it so happens, had said nothing of the sort, anymore than he had said that

> it is a custom [in Ireland] that the poor who stand at churches to beg for alms, even though they are naked, should you, in lieu of alms, give them a piece of stone, which they are wont to burn instead of wood, pay thanks to you for it and go off all cheerful and rejoicing.

Pius had indeed mentioned such a 'miracle', but according to what it said in *De Europa*, which was clearly Bordone's source on the matter, it was a miracle that occurred not in Ireland but in Scotland, and unless Bordone had consulted a corrupt edition of the work, which is a possibility, all we can conclude is that his attitude to the books he talked to was cavalier, to say the least.[12] Still, it was presumably poetic justice that Ireland, which had been deprived by Pius of its barnacles, should in return receive an even more amazing wonder: beggars who are not ashamed to stand stark naked at church gates and who show joy and cheer at being given stones to burn.

Perhaps Bordone needed such a miracle in Ireland to give the lie to those who still confused the island with the Island of the Blessed. 'In the part which lies towards the west', he informs his readers, Ireland 'has a gulf in which, according to the moderns, there are three hundred and sixty-eight islands which are called blessed, fortunate and holy'. That is what they may be called, he seems to want to indicate, but how could a land be deemed 'fortunate and holy', when its idea of holiness is to undress before the mansion of the Lord (Pius, it should be noted, had said that the beggars were 'almost' naked), its idea of good cheer to receive a stone, and its idea of honour to eat dead relatives? The editor of Peter Martyr was to get it right, for sure. The kind of savagery which Ireland stood for was not the savagery of the noble kind: nobility, in the world's 'noblest part', was of a different order, and blessedness far less fabulous. And yet, dismissing what 'the moderns' ascertained was not as easy as all that. The moderns, it is true, were always to be favoured above the ancients, at least explicitly.

Thus Ptolemy, for instance, is never given the benefit of the doubt in the way which Pius is. Where Ptolemy had said that Ireland was two hundred and forty miles long, Bordone explains that, 'according to our times', its length is four hundred and fifty miles and that, where Ptolemy had called the nearest point to Britain 'Ismanium', in the vernacular the place is now called 'Cenofrit'.[13] But could it be that one day the moderns too would be discovered to have erred?

The nagging doubt, apparently, accompanied Bordone until the very end, and when all was said and done, he could think of no better way of dispelling it than laying the blame for it on others. To defend himself against accusations of having been economical, or spendthrift, with the truth – whether in terms of Irish wild geese or of other fables of the sort – he accused his would-be accusers of setting too much store by the truth. 'There may be some', he wrote in the conclusion,

> who will say that in writing of these islands it would have been enough only to have told of their location and their circuit and that, as I wrote of them, I need not have bothered to tell the fables and the stories that happened on them, since such things are what are shown to children who have only just learnt to read. To those I answer that, even though someone may already know something, to be reminded of it again is nonetheless often the cause of pleasure, but should we accept that those who remember it could find it tedious, then they should let those read of it who know nothing at all about it.[14]

That is certainly a turn-about. Having promised nothing but the truth, when all is said and done, Bordone has no option but to reveal his hand as an Ariosto. Let those who wish upturn the story – he signals to his readers – but leave me with my fables! Is not the truth *for* more important than the truth *about*? And what is wrong with holding on to an *Ibernia* that is *fabulosa*?

* * *

Not everyone, of course, believed in holding on to an *Ibernia* of the sort. Virgili, as we remember, was adamant that it was 'futile to discuss' the miracles of Ireland 'once again', and Virgili was to have an influential disciple in Tommaso Porcacchi (1530–85), the Italian who succeeded Bordone as the self-proclaimed 'connoisseur' of islands and who, paraphrasing Virgili, proclaimed in the island-book he brought out in 1572: 'as for those many miracles which are said to be in Ireland, given that, the more you look for them, the less you tend to find them, it is vain and unwise to record them'.[15] Porcacchi, like Bordone, was a market-savvy writer who lived by his pen and worked all his life for Venice's publishing industry. Besides authoring works of his own (such as the island-book), he translated the works of others (Pomponius Mela's *Cosmography*, for instance) and produced annotated editions of works that were becoming classics (like *Orlando Furioso*); and all of these enjoyed great popularity, not least *The Most Famous Islands of the World* [*L'isole più famose del mondo*], which became the new reference work on the subject, going through no fewer than six editions in the course of the century following its initial publication, notwithstanding the fact that Porcacchi was no less of an armchair traveller than Bordone and played no lesser tricks with the truth. His description of Ireland, indeed, is not so much a paraphrase of Virgili's as it is a plagiarism. It is in Italian, but its contents are Virgili's, almost to the letter, and its true authorship is not acknowledged.

Naturally, being a translation it could not remain faithful to the original in every detail, but not all the changes which Porcacchi introduced – and he did introduce a good few – can be explained in terms of translation's inevitable losses, any more than they can in terms of Porcacchi's rush to satisfy the needs of the market. In some instances Porcacchi clearly had difficulty in understanding what Virgili meant. For example, instead of distinguishing, as Virgili does, between 'the more polished Irish' and 'the backwoods Irish', he talks of 'city-dwellers' and 'peasants'. Where Virgili had linked the seditiousness of the Irish to their propensity 'to seize power', he links it, surprisingly, to their propensity 'to obey'. Even more surprisingly, he misconstrues Henry II's reasons for summoning the bishops and the 'kinglets' of the island (whom he in fact refers to as 'petty lords') in order – as Virgili had said – to 'discuss the question of how to render dominion over the island permanent for himself and his heirs'. According to Porcacchi's version, it is the rule of the petty lords that is 'to last forever'.

In other cases Porcacchi's changes seem designed to facilitate his readers' comprehension. Thus, in connection to what was referred to in a previous chapter as the Irish custom of dye-cleaning their clothes with saffron, he specifies that they do so 'in order that the stains on them could not be seen'. And, subtracting rather than adding, he decides not to translate the sentences in which Virgili had explained that among the earlier inhabitants of the island there had been the Scots, who had come from Scythia, or possibly from Egypt, but who had then joined forces with their cousins the Picts to settle the top-most part of Britain, which therefore became Scotland, while 'bit by bit Ireland got back its original name'. He must have reckoned that his readers would have been confused enough to find out that the Irish had originally been Spanish, and all the more so as he had taken it upon himself to mention that Hibernia 'today, by what derivation I do not know, is called Ireland [*Irlanda*]'.[16]

Porcacchi also appears, at first sight, to have taken it upon himself to alter, by omission or commission, the ideological charge which Virgili had given to his description. As has already been mentioned, those Virgili had deliberately called Ireland's 'kinglets' [*reguli*] Porcacchi calls its 'petty lords' [*signorotti*]. At the same time, although he makes it clear that Irish 'cities, because they obey the English king, have more honourable manners', he chooses not to mention that they are 'least uncivilized'. Neither does he make any mention of the meat which Virgili had been at pains to introduce into the Irish diet, thus rendering it impossible for his readers to pick up the intended reference to Caesar and his enemies. On the other hand, Porcacchi does not simply say that 'in their beastliness, the Irish practice the Christian religion chastely', but 'in their beastliness and wildness'. True, the latter may be no more than a case of using two words instead of one for stylistic effect, as happens so often in Italian; and the other inclusions or exclusions could also be the result of stylistic considerations, or more likely, of financial ones. Then, like now, publishers' deadlines must have meant that effect and fidelity in translation could not but play second fiddle to efficiency and punctuality. Yet, for all these changes, it is clear that overall Porcacchi was well aware of the general thrust of Virgili's description and had no intention of undermining it – quite the contrary. So as to re-establish the preferred balance between chorography and chronology, he concludes the description with what in Virgili had been its introduction – that is, with Henry II's conquest and with the pope's concession of Ireland to that 'most mighty and good Christian king' ('good', it should be noted: Virgili had simply said 'most mighty') – and thereupon adds a telling final sentence of his own: 'these people then rebelled on other occasions against the kings of England, but they were always defeated, and so now they also obey the English crown'.[17] Obviously the prerogatives of empire were too well rooted in the minds of geographers to be upturned. No rebelling and no writing of the world could be allowed to stand in the way of its forward march, even though the focus on the 'done' and 'to-be-done' of history had lost its sharpness.

* * *

What could stand in the way of empire, however, were the forces of the market and of religion, which together could play havoc not only with imperial dreams but also with Hibernian fables. When Porcacchi wrote his island-book, the world, old and new, was firmly in the grip of the religious conflict unleashed by the Protestant Reformation and fanned by the Catholic Counter-Reformation, but Porcacchi, at least as far as Ireland was concerned, decided to ignore it. That was the price to pay, apparently, for plagiarism. Yet plagiarism could come in many guises, and if the need arose there was no reason why it could not take account of the conflicting forces which tore the world apart even as, paradoxically, they brought it closer together. An intriguing case in point, in the third category of writings of the world mentioned above, namely cosmographies, is to be found in the *Cosmography or Description of Every Land* [*Cosmographey oder Beschreibung aller Länder*] by the German-Swiss author Sebastian Münster (1488–1552), a Franciscan monk turned Protestant, printer and professor, who is famous today primarily for his achievements in the study of Hebrew and the printing of Hebrew texts, but was particularly famed at the time as the German Strabo, because of his special interest in geography. He edited and published the works of Solinus, Pomponius Mela and Ptolemy, but above all he wrote the massive *Cosmography*, which was so well known and so widely read, in both Protestant and Catholic lands, that it has earned the reputation of having been the 'lay Bible' of sixteenth-century Europe. Between 1544, when it first saw the light of day (in German) and the middle of the following century, it went through no fewer than forty-six editions, in six different languages, including Latin and Italian.[18] Strictly speaking it is a work which does not fall within the purview of this book, since its author was not Italian and most of its editions were published north of the Alps, the vast majority in Basel; but given its popularity, and the fact that it was translated into the two *linguae francae* of Italy's élite, it would be a mistake to ignore it, especially as much of what it says about Ireland is actually of Italian derivation.

The textual history of the work, which grew out of the German original, which itself had grown out of an earlier work of Münster's entitled *Map of Europe* [*Mappa Europae*], is extremely complex, not least because the text continued to be added to by Münster's heirs; but it need not detain us here. Instead we will focus solely on the Italian editions (the first of which was published in Basel, posthumously, in 1558, the second in Cologne in 1575, and the third in Venice, also possibly in 1575) together with the Latin editions on which they are modelled (the first of which was published in Basel in 1550).[19] The *Cosmography* actually deals with Ireland in two separate chapters: briefly in a chapter on the 'British Isles', which is little more than a compendium of data drawn from Ptolemy, and at greater length in a chapter entitled 'New Hibernia', and it is this later chapter which deserves particular attention. What makes Muenster's New Hibernia new is essentially that it is not derived from Ptolemy. To us, however, it is all old hat, since it is in fact a *mélange* of sections from Virgili and from Giovio, interspersed with passages from Solinus, and with only very little which is truly original and new. In its earliest phase it contains more of Virgili than it does of Giovio, but later on it is Giovio who takes over, although Virgili still continues to be present. It is thus a mix of oil and water, so to speak, and Muenster and his heirs appear to have realized as much, as we will see. Yet presumably they trusted no-one would notice. They should probably be compared to computer-age students, who cut and paste from Internet with abandon, in the hope that no-one will mind and the marks of their assignments will not be affected.

The Irish cut-and-paste job carried out by Münster and his heirs was done as follows. They did not borrow from Virgili any of those elements which are the prime conveyors of its ideological charge. There is thus no reference in their *Cosmography* either to the way in which the Irish fight, or to the way in which they eat, or talk, or dress and wear their hair. Their kinglets are not mentioned, and neither are their cities, or their education, or their funerals. They are neither said to be lazy, nor faithless, nor sober, nor beastly Christians. More importantly, no allusion is made to the distinction between the 'backwoods' Irish [*silvestres*] and the 'more civil' ones [*civiliores*]. True, the Irish are said not to want to spare much time for agriculture, and to be 'inhospitable, wild and cruel' and therefore to be good at fighting wars, which is what they deem to be most honourable, but generally speaking what is retained from the *Anglica historia* is essentially the geographical features of the description: Ireland's name, its position, shape, size and orientation, the names and location of the four parts into which it is divided, the fauna it has and does not have, and the excellence of its climate and fertility of its soil.[20]

By the same token the ideological force was also drained from the description of Ireland by Giovio, at least initially (i.e. in the 1550 Latin edition and the 1558 Italian one). In fact, all that is left of that force is *Connatro Honel*, 'a man illustrious in arms', who in this case has been demoted to being a mere 'governor' [*governatore*] of Ulster and who does little more than lead a charge of 4,000 horsemen and 12,000 foot-soldiers against 'Howard, king of England' and then make peace with the English, thereafter to 'live quietly'. More obligingly, he then promptly passes away – long before his actual death – so as to allow the English to focus all their energies instead on 'Captain' Silken Thomas:

> when he had died, Thomas Fitzgerald [*Thomaso Giraldino*], the Captain of Hibernia, remembering that his father had been killed by the English, caused a sedition, in about 1534, and mustered more than forty thousand troops, both foot soldiers and horsemen; but the revolt was then put down, at great danger and harm to the Irish, and with certain terms attached.

In the later Latin versions of the *Cosmography* (but not in the Italian ones, both of which may be pirated editions and seem to remain unchanged) Conn, although he still dies a premature death and is still made to fight against 'Howard king of England', then leads a second charge of the 4,000 and 12,000 (against plain Thomas Howard this time) and once again makes peace with the English (on slightly better terms), as almost everything from Giovio's description that had originally been left out is now included in amongst Virgili's (unchanged) words.[21]

What is more, in this later version of the description, to the information which had originally been provided regarding St Patrick's mission to the Irish in AD 433, is also added the information, somewhat confusingly, that the Irish first received the Christian faith in AD 360. In addition a list is provided of Ireland's four archbishoprics and fifty-two bishoprics – some with names that sound distinctly odd: *Damligiorem* under the Archbishop of Armagh for instance, or *Furichland* under the Archbishop of Dublin, *Gogmogha* under the Archbishop of Cashel, or *Delundunuch* under the Archbishop of Tuam – while its bishops and its churches are made to feel the wind of more recent confessional upheavals. Giovio, it will be remembered, had written that 'all the Irish, even in harsh and wooded [*silvestres*] places, observe the Christian rites. They are taught by teachers appointed by the bishops for their suitability as masters of Latin and music'. Muenster's heirs, however, chose to say instead: 'their places of worship are mostly situated in woods [*silvis*] or on elevated sites, and they have learned bishops, who are able to sing and teach, especially in the Reformed churches'. It was not just Protestants, though, who needed to be kept on side, and so it is we now find out that Ireland's hobby horses, which have ceased to march in papal processions, are gifted not only to women of the better sort in France and in Italy, but also to German 'princesses'. At the same time we are invited to assist at the return, on the back of Giovio's text and for all Virgili's (unrecorded) disapproval, of Ireland's marvels, which, however, are no longer quite what they once used to be. The piece of wood that is dug into the magic lake simply changes into iron, and the barnacles – whatever Messer Andrea Rossi might have had to say on the matter – are now called *claccos*.[22]

How typical the description of Ireland is of the *Cosmography* as a whole and whether other countries were subjected to a similar makeover by Münster and his heirs, it is hard to tell. What is for sure is that the most fabulous aspect of the Münster description is not so much the transformation of Ireland's fables as the emasculation, so to speak, of its history, or rather of its chronology, for the benefit of its chorography. A wisecrack might be inclined to say that the result is a typical product of Swiss neutrality, neither for *imperium* nor against it, quite the opposite. Certainly the gamble – if that is what it was – paid off and the work, as we said, enjoyed enormous popularity, on both sides of Europe's imperial and confessional divide.

* * *

The divide, however, could not be ignored and cosmography would soon be made to toe the line. The first major Italian cosmographer of the sixteenth century was Giovanni Lorenzo d'Anania from Calabria (c.1544–c.1608), who was 'born of noble and Catholic parents' and who – or so at least a friend of his would have us believe – was predestined to write the world. When he emerged from his mother's womb – the friend informs us – d'Anania slipped through the fingers of the midwife and, falling to the ground, he opened up his 'little arms' as though to kiss the earth, and in so doing he gave an unmistakable sign of 'the marvellous study he was to make of it, which is clearly manifest in his work on *The Entire Edifice of the World* [*Universal fabrica del mondo*]'. Predestination, though, was not a guarantee of omniscience and it was to be another thirty years before that 'marvellous study' saw the light of day, during which d'Anania, yet another armchair traveller, engaged in a great deal of (one-way) talking with books. The books he talked with are listed in a lengthy appendix to the *Edifice*, and they include Bordone, Fazio degli Uberti, Paolo Giovio, 'Giraldo Cambriense d'Hirlanda', Isidore of Seville, 'Pope Pius', Pomponius Mela, Tommaso Porcacchi, Ptolemy, and Solinus. But there is no mention of Virgili, for Catholic *noblesse*, it would seem, could also come to terms with plagiarism. Equally it could not but acknowledge that predestination, no more than it was a guarantee of omniscience, was a gage of infallibility. After the *Edifice* was first published, in Naples in 1573, it was corrected and augmented, before being issued for a second time, in Venice in 1576, and thereafter for another two (Venice, 1582, 1596). The result, as far as Ireland is concerned, is a cut-and-paste job not unlike the one which had seen the light of day in Protestant Basel, although admittedly it is held together by a greater sense of purpose. The edifice of the world, in d'Anania's estimation, is built according to God's plan – and of course, according to the plan of the ancients too.[23]

It begins with 'the island of Ireland [*Hirlanda*], which the inhabitants now call *Erin*', because that is where Europe begins and Europe is the world's 'happy part' [*felice parte*]; and although the inhabitants of Ireland are not explicitly said to be a 'happy people' in the manner of Petrarch, d'Anania does most certainly paint the place as a blessed land and a land of plenty. It is devoid of snakes and it produces grain and silver in abundance, as well as fish, iron, wood, leather and butter, not to mention falcons, hobby horses, dogs and eagles; its pastures are so rich that its cattle is at risk of over-grazing if it is not led away before the day is out; but best of all, its climate is so 'marvellously' healthy that in the second and later editions of the work d'Anania no longer deemed it good enough to say that it was 'temperate' and chose instead to endow it with an almost sacramental quality, 'goodness' [*bontà*]. It is in its miracles, however, that Ireland's special status is made manifest, and though to some they may appear to defy credulity, it is nevertheless a fact (based, we must presume, on 'Giraldo Cambriense d'Hirlanda') that they were brought about 'by a secret necessity of divine providence, by which reason's eyes, since they are not natural things, are blinded, as is a night bird by the rays of the sun'. Their number is thus increased from three to eight between the first and subsequent editions, and the later readers of the work are informed that in Ireland there are rivers which change wood to rocks and stone to iron, an island in Lough *Foilo* in Ulster [*Hultonia*] where people cannot die and to do so must be brought elsewhere, another island where no animal can live, a spring in Munster [*Momonia*] which turns the hair of those who wash in it to grey, and one in Ulster which has the opposite effect, the Isles of Aran where corpses do not decompose and so are left to stand in people's gardens, and Lough *Erno* which used to be a spring but flooded and became a lake because of the 'criminal bestiality' of the local inhabitants.[24]

Also included among Ireland's miracles is St Patrick's Purgatory, which the strictures of the Counter-Reformation presumably suggested to d'Anania was best reconfigured as a tourists' rather than a pilgrims' destination – of the sort, we might add, which would have pleased Guerrino il Meschino. It consists of 'nine caves', which one visits 'out of human curiosity' and in which one is subjected to visions of the pains of hell which cause 'fright and terror'. Elsewhere too tourists would have had good reason to take fright, for the Irish who live inland are 'half wild, rough and cruel-minded' and 'in their speech are totally barbarian'. Nevertheless, just as Chiericati had done, visitors to Ireland would have taken comfort in the knowledge that the Irish were uncommonly devoted to the Church (the right kind of church, of course):

> They have two metropolitans, one in Armagh and one in Cashel, who, as in the east, are normally for the most part monks, and for whom they have so much respect that foreigners can walk anywhere under their guidance, for all that the Irish are much given to robbery. They devote a great deal of time and effort to the study of Scholastic theology, which is the reason why many of them have remained Catholic until today.

In other words, though the (Catholic) visitors – 'by a secret necessity of divine providence' perhaps – may have felt secure in Ireland under the 'guidance' of the Church, we cannot help but notice that they might equally have felt unsure, like Chiericati, as to the spirit in which to take the place, and indeed, whether they might not have felt that the proper spirit was in fact Virgili's. For all that the Irish 'have remained Catholic until today', it is nevertheless an oxymoron which best sums them up. They are 'beastly Christians', Virgili had said. They are 'Scholastic robbers', says d'Anania.[25]

That is not the only echo of Virgili in the Ireland of the *Edifice*. Like Virgili d'Anania has problems with Ireland's ecclesiastical provinces, and he does not seem to know what to make of its towns and cities. Between the first and the second edition he reduces the number of Ireland's archdioceses from four to two, and its suffragans to none at all from 'many', and he deletes a good few names from the extensive list of place-names he had initially inserted, thus depriving later readers of the work of places such as *Derra* [Derry], *Molengara* [Mullingar], *Novan* [Navan], *Roscaman* [Roscommon], *Sligaha* [Sligo] and *Trahila* [Tralee] among others. The way the Irish fight is also treated ambiguously. D'Anania recognizes that they are 'valorous and trained for war' and, Giovio-like, that they 'are so enamoured of liberty, that they are happier to die than to endure servitude'. At the same time he suggests that if some of them have managed to hold on to their liberty, it is probably due less to their valour and their military training than it is to the marshes, lakes and woods which serve them as fortresses and castles. In any case they have now shaved off their moustaches styled to give fright to the enemy (although they still have a preference for long hair) and Irish noblemen now mostly pass their time in playing music and in hunting, and only go to war 'occasionally'.[26] But that is probably for the best.

When reporting that until quite recently the Irish had their own king – whose royal seat was in Armagh and who claimed to hold his power from the pope – but that their country is now 'under the dominion of the English', d'Anania does not show the slightest sign of regret, as Giovio might have done. What is more he mentions with approval the fact that the Irish who live along the coast, 'thanks to their intercourse with foreigners, are civilized and friendly'. He even seems determined to precipitate the demise of native Irish customs. Remarking that if they live inland, the Irish normally wear 'clothes of yellow cloth, in the gypsy style', but if they live along the coast, they dress 'almost in the English style', from the second edition onwards he deprives them of the means to dress like gypsies. Twice in the first edition he had mentioned that Ireland produces vast quantities of saffron, but in the second and the later editions he takes out the saffron and replaces it with fish and iron. Better English than gypsies – would seem to be the message. And if we find the message surprising, we need but think of Berlinghieri. For Berlinghieri, we will remember, any *imperium*, even the Turkish one, was better than no *imperium*, and writing the world was inexorably part of the civilizing mission which sought to fulfil *imperium*'s destiny, even at the expense of the 'Gherardine empire'. The new Turks were the Protestants, but *imperium*'s prerogatives remained the same. Porcacchi recognized it, and so did d'Anania.[27]

* * *

So too did Giovanni Botero (1544–1617), a one-time Jesuit who, having been honourably discharged from the Society, became secretary to the archbishops of Milan, Carlo (1538–84) and Federico Borromeo (1564–1631), and subsequently to Duke Carlo Emanuele I of Savoy (1562–80–1630), and who, besides acquiring fame as the major theorist of *raison d'état*, was also the last great Italian cosmographer of the sixteenth century, authoring a work, the *Universal Reports* [*Relationi universali*], which became so widely read – with no fewer than twenty-three editions in the original Italian, as well as seven English editions or reductions, five Latin editions, three Spanish ones and two German ones published between 1591 and the middle of the following century – that it was to serve, or so it has been claimed, as 'the geopolitical manual of all of Europe's ruling class'.[28] Whether it did or not, there is no doubt that Botero's is a geography which is aggressively militant. Its aim, on the one hand, was to ensure that the 'noble minds' to whom it was addressed behaved like responsible adults and did not shield themselves from the rest of the world, for people, says Botero (as though addressing himself to the Countess of Pandino and her friends), 'who know no other part of it except for the place where they were born, and whose thoughts are limited by the extent of their lands, and whose mind goes no further than their villas' are no better than small children. On the other hand, it aimed to subject the world to a single jurisdiction. In this case, however, the confession of that jurisdiction was not a matter of indifference, for the rulers to whom Botero addressed himself were above all the princes of the Church, whose duty it was 'to propagate the faith with all their might', since

> the power of the Roman Church has no limit, the jurisdiction of the Holy See no end: it passes over mountains, crosses the ocean, embraces the one and the other hemisphere, and has as its mission to lead all peoples to God, and as its goal to raise the earth to heaven.

For that reason, the *Universal Reports* seek above all to 'describe the state in which the Christian religion finds itself today throughout the world'.[29]

Botero's Ireland is thus the stage of a far more overt religious conflict than had been the case hitherto. Yet still it is the locus of an oxymoron:

> The most inhabited part of this island is the eastern part, and the southern part up to the borders of Munster [*Mononia*], in which area there are two archbishoprics, Armagh [*Armagnac*], which is the metropolis, and Cashel [*Cassel*]. But the most important and noblest city is Dublin, with an English garrison, beneath which lie the more civilized and inhabited parts. But the other parts, which are towards the north and west – Ulster [*Ultonia*], Connaught [*Connachia*], Munster – obey different lords, who are protected by marshes and lakes and woods, which are their ditches and their ramparts and their castles, but more so by their poverty, since there is nothing to gain from them; and they are safer in summer than they are in winter, because the water, which freezes in winter, in summer floods the fields. And these more than the others have remained true to the Catholic faith, against the machinations and the violence used by the English to infect them with the evil of Calvin and of Zwingli. The inhabitants, who, as Strabo writes, used to consider it praiseworthy to eat their dead parents, are still today quite boorish and wild. They are enemies of toil, and are

> of little worth. It is true that those who live along the coast, because they mix with foreigners, are friendly and civilized, up to a point.[30]

The Irish, that is, for all that they are 'true to the Catholic faith', are also true to tradition: they are 'boorish and wild' and 'of little worth'. For their part, the English, even though they engage in 'evil machinations', are still the more civilized.

In many other ways too Botero's Irish are faithful to tradition, so much so that one cannot help but wonder whether it was not he whose mind remained enclosed within the walls of a villa, since all he did, apparently, was copy down what others said. Taking his cue from Boccaccio, by the looks of it, he first of all refers to Ireland's mountains, woods, marshes, ponds and lakes, and then, as though nodding at Petrarch, he lists the animals it has and does not have, and highlights the absence of earthquakes and of thunder. Later on it is with Giovio-like approval that he mentions the enormous quantities of fish the Shannon and the Irish Sea produce, and at one point he appears to follow Andrea da Barberino when he suggests that Ireland is a fitter place for bovine than for human habitation. It is Virgili, though, who makes his presence felt like no-one else, for all Botero's indifference to Virgili's dietary prescriptions (the Irish, Botero says, only feed on milk and honey) and to his recipe for dye-cleaning (Ireland's sheep are black, Botero also says, and so the Irish do not need to dye their cloth). However, it is undoubtedly Virgili whose oxymoron lies behind Botero's. To claim the Irish were 'true boorish Catholics' was little different, all told, from saying they were 'beastly Christians'.

In one significant respect, however, Botero does take his distance from Virgili, and he does so with the help of what may seem like a most unlikely ally, given his commitment to the truth and nothing but the truth. 'Of this island things are told', he writes, 'because of which Ariosto called it fabulous', and thereupon he lists a good few Irish 'fables', mostly magic springs and magic islands, some of which are explicitly attributed to 'Giraldus Cambrensis', and one of which in particular catches his attention: an island in Lough *Dere* in Ulster which is divided into two parts, one pleasant and charming, the other, where there are nine pits, unpleasant and frightening, and 'whoever chances there and sleeps the night is tormented greatly by wicked spirits. This place is called St Patrick's Purgatory'. Ariosto, it would seem, provided just what the Counter-Reformation needed. 'Wickedness' and 'torments' have been added, but Ireland's 'little old saint' is worthy of no more than a 'chance' encounter. Gone for sure are the days of Guerrino il Meschino, when the faith could not have been propagated *but* for St Patrick and his Purgatory.

Yet Ariosto plays an even greater part than that in Botero's universal scheme of things. The *Reports*, as we know, are for 'noble minds', but noble minds, it is clear, cannot feed on *utile* alone. They also need their *dulce*, and Botero therefore never fails to point out how much 'delight' the universe can offer. Nevertheless he was not entirely confident, apparently, that to some of them that may not be enough. The sheer size of his *Reports*, he seemed to fear, may put them off. What possible delight – they might ask themselves – could be derived from reading such a massive work? Botero therefore tried to reassure them. I promise you – he wrote in the dedication of the section on islands in a later edition of work – that if you read my book after a hard day's work, you will be able 'to find shelter in famous England or in fabulous Hibernia'.[31] Put otherwise: what are you waiting for? Go there, and you will find there is no better *otium* than fabulous Hibernia's. There was no point, for sure, in causing fright to 'noble minds'. They needed 'shelter', and if they did not like the newness of the world, *Ibernia fabulosa* was there to help them turn it upside down. Why should they bother with the truth, if the fables told a better story? Let there be a new world, let empire have its demands, let the faith be propagated, let geography be militant, but ... let the readers not be surprised!

Notes

1. '... monti altissimi [...]. E' opinion che non abbino determinato parlare fra loro come han tutti gli altri uomini del mondo; e che non sappino ciò che sia signore [...] ma che sian del tutto salvatichi animali, eccetto che hanno l'effigie umana' (MARTIRE (1985), p. 194). On Martire see Stoppa and Cicala (1992).

2. '… grande e terribile […]. Né si maraviglino li lettori che in questa isola tanto lontana da noi si truovi questa generazione d'uomini salvatichi, che ancora nell'isola Hibernia, qual è sotto il re d'Inghilterra, non troppo lontana da quella, nella parte fra terra, dove non è altro che selve e monti altissimi, si sa trovarsi uomini infiniti salvatichi, quali mai hanno voluto aver commerzio con quelli che abitano appresso il mare, né si son potuti mai espugnar dalle genti del detto re' (Martire (1985), pp. 194–5).
3. 'La Ibernia oggi chiamata Irlanda e posta fra l'Inghilterra et la Spagna, et tira di lunghezza 280 miglia in circa et di larghezza 100. Ha 50 fra Vescovati et terre, li principali fiumi suoi sono Suiro, Boanda et Sineo, quali si dividano in quattro parti cioe Momonia, Laginia, Hultonia et Connaccia. In questa Isola e tale temperie di aria che non solamente non vi nasscie animale alchuno venenoso ne erba, ma ancora, se vi e porto d'altronde non vi aligna ma si secca o muore. Nella parte settentrionale di questa isola e una profondissima voraggine, laquale alcuni gravi scriptori dicano essere il pozzo di S. Patricio vescovo di detta isola. Parimente affermano che in una delle isole orcade e uno arboro il quale stando con le radici nel aqua et nuota atorno il lito li cui pomi cadendo nel aqua producano Ucelli' (Firenze, Palazzo Vecchio, Guardaroba Nuova).
4. '… l'Hibernia hoggi detta Irlanda, et da gli Inglesi Irlond […]. Cesare nelli suoi comentarij dice detta isola essere in giro 2000 miglia et il venerabil Beda 2600, ma li moderni (ali quali piu si deve credere secondo il precetto di Ptolomeo) affermano non essere piu di 1700 […]. La gente di questa isola che dagli antichi fu detta non havere lettere, ne musica hoggi nell'una et l'altra faculta si vede essere eccellentissima et sopra il tutto valorosa in arme, et molto intenta ala professione della nobilta et dello splendore, ma hoggi sono tutti eratici et segreghati dala Chiesa Catholica' (Firenze, Palazzo Vecchio, Guardaroba Nuova). For more on Danti's maps see Fiorani (2005), especially Chapter 2 ('The Guardaroba Nuova', pp. 17–32) and Chapter 7 ('The gallery of maps', pp. 171–207).
5. 'Con le lor nauigazioni non inuestigorono piu oltre che quello che da gli loro antichi ritrouorono scritto' – 'quelle che ne tempi nostri si sono ritrouate' – 'gl'huomini de' tempi nostri, che con grandissima perdita della lor faculta, & etiamdio della lor propria uita, non hanno di ricercare il mondo in ogni parte mancato, il circoito della terra inuestigando' – 'molti errori che gl'antichi a posteri lasciato haueuano' – 'insegnare a quelli che non sano' – 'quelli, che d'altro studio occupati si trouano, & anchora quelli che al nauigare inchinati non sono' – 'tutte l'isole, & penisule del mondo con lor nomi antichi & moderni, & con ogni altra cosa che a quelle s'appertengono io faro intendere, si delle istorie che de quelle scritte sono, come etiandio delle lor fauole, & in qual parte del mare giacciono, & de uarii costumi che tutto di nauigando ui si ueggono, & sotto qual parallelo, & in qual clima siano poste' (Bordone (1965), dedication).

6. On Bordone see Armstrong (1996), Billanovich (1968), Donattini's introduction to Bordone (1982), Guglielminetti (1994). The manuscript illuminated by Bordone is the *Evangeliary of Santa Giustina*, dating from 1523–5 (Chester Beatty W 107). It can be partially viewed at http://www.cbl.ie/cbl_image_gallery/collection/detail.aspx?imageId=152&ImageNumber=T0000630&collectionId=3&page=0.
7. 'Occeano orientale & mar indico' (Bordone (1965), title of Book Three). 'Dovendo io dvnque nipote mio carissimo dar principio a questo ragionamento delle isole del mondo, me par cosa conueniente incommenciare da l'uno de capi del nostro continente (benche piu picccola la europa de l'altre due parti sia) nondimeno di forza, ingegno & sapientia sempre l'altre due parti ha soprauanzato, & per cio, da questa parte, si come dalla piu nobile del mondo pigliero il principio mio' (Bordone (1965), p. Ir).
8. 'ottimo conoscitor' – 'eccellente cirvgico' – 'sopra le potenti armate de segnori Venetiani, & del chatolico re [...] nauigando tutto il mar mediterraneo' – 'dependendo l'Astrologia da cieli, consequentementte e cosa perfetta, donque perche la Cosmographia dipende da l'Astrologia, e cosa uera' (Bordone (1965), dedication).
9. 'Nella quale scotia, secondo che recita Pio nel suo itinerario, ui sono arbori (che presso le riue di uno fiume notano) & producono frutto allo aneto simile, liquali, come sono presto che maturi, per se stessi caggiono, parte in acqua & parte in terra, quelli che caggiono in terra diuengono putridi & marci, ma quelli che ne l'acqua caggiono, diuentano uccelli pennuti, che poi che sono fatti grandi, per l'aria come gl'altri uccelli uolano, della qual cosa piu diligentemente inuestiga[n]do cotal cosa, fu certificato non ne la scotia, ma nelle isole orcade, esser cotal miracolo' (Bordone (1965), pp. IIr–v). 'Didicimus miracula semper remotius fugere famosamque arborem non in Scotia, sed apud Orchades insulas inueniri' (Pius II (2001), p. 186).
10. 'Chi non sa? che a chi non hauesse ueduto uno etiopo non agieuolmente ui si gli darebbe a credere che uno huomo fosse nero? Ma molte uolte la natura produce cose, che paiono impossibili, & nondimeno pur sono, cosa non così ageuole e da credere, che le frondi de alcuni arbori lequali caggiono nelle acque, diuengano uccelli pennuti, & questi pur si puono in Vinegia uedere appo messer Andrea rossi, che de hispagna, gli fece portare, liquali sono minori delle oche, & maggiori della anitra, & sono da hispagnoli appellate grauagne' (Bordone (1965), pp. LXXIIIr–v).
11. 'Oltre a questo non ui e cosa, che di memoria degna sia, saluo che, questi isolani sono grandissimi mangiatori, & tra loro cibi la carne humana, hanno molto in uso, Et li loro parenti poi che morti sono, si mangiano. Et questo appo loro e

grandissimo honore, ne meno di questo apprezzano con le loro femine in publico, & con madri & sorelle mescolarsi' (BORDONE (1965), p. Iv).

12. 'Questa isola è piana, bene habitata, & gl'habitanti molto piu che li britani del saluatico tengono, non dimeno sono buoni mercatanti, non produce olio, uino ne grano, beuono ceruosa, usano pane di farina di orzo & di segalla, è abondante di fiumi, ma piccoli secondo che dice Pio. [...] Et una cotale usanza tiene che li poueri, che stanno alle chiese per dimandar limosina (benche nudi siano) se tu lor doni un pezzo di pietra, laquale hanno in uso d'ardere in luogo di legna per limosina, te ne rendono gratie & se ne uanno tutti lieti & festanti' (BORDONE (1965), p. Iv). 'Illud tamen nobis in Scotia miraculum representatum est; nam pauperes pene nudos ad ecclesias mendicantes acceptis lapidibus, elimosine gratia datis, letos abisse conspeximus. id genus lapidis siue sulphurea, siue alia pingui materia preditum pro ligno, quo regio nuda est, comburitur' (PIUS II (2001), p. 186).
13. 'Alla parte, che uerso ponente giace, ha uno golfo nel quale ci sono (secondo moderni) isole trecento sessantotto, lequali sono dette, beate, fortunate, & sante' – 'secondo i tempi nostri' (BORDONE (1965), p. Iv).
14. 'Saranno forsi alchuni che diranno che nello scriuere queste Isole: bastaua solamente di hauer narato il loro sito: & circoito: senza hauermi nello scriuer faticato di dir fauole, & historie sopra quelle interuenute: conciosia cosa che a fanciulli nelle prime littere, loro dimostrate siano a quelli respondo, che quantunque alchuno sappia alchuna cosa, nondimeno il piu delle uolte, gli piace quella di nuouo sentire ricordare, ma pogniamo che quelli che nella memoria le tengono, a noglia gli fussero, di leggere le lasceranno a quelli che del tutto non le sanno' (BORDONE (1965), conclusion to Book Three).
15. 'Quei tanti miracoli poi, che si contano esser nell'Irlanda, percioche quanto piu se ne cerca, meno se ne suol trovare, è vanità & poca prudentia ricordargli, quando io attendo non alle favole ma all'historia' (PORCACCHI (1572), p. 72). See VIRGILI (2009), pp. 234–5 (25). On Porcacchi see Guglielminetti (1994), Lastraioli (2006).
16. 'Cosi cittadini, come contadini' – 'essendo avezza alle fattioni [...] è [...] presta [...] a ubidire' – 'havendo [...] convocati i Vescovi e i Signorotti per trattar del governo loro, c'havesse havuto a durar in perpetuo' – 'accioche non si veggano sopra le brutture' – 'Hoggi non so con qual derivatione vien detta Irlanda' (PORCACCHI (1572), pp. 70–2). See VIRGILI (2009), pp. 220–1 (1), 224–5 (11), 230–1 (17).
17. 'Le città nondimeno, come quelle ch'ubidiscono al Re d'Inghilterra, hanno costumi piu honorati' – 'in questa lor bestialità, o fierezza osservano castamente la Relligion Christiana' – 'un potentissimo e buon Re Christiano' – 'Si son poi questi popoli altre volte ribellati da' Re d'Inghilterra: ma però sempre sono stati

vinti, & cosi hora ubidiscono similmente a quella corona' (PORCACCHI (1572), pp. 71–2). See VIRGILI (2009), pp. 220–1 (3), 226–7 (15), 230–1 (16).

18. On Münster see Broc (1996: 'German Strabo', p. 66; 'lay Bible', p. 72), Burmeister (1964), Hantzsch (1965), McLean (2007), Skelton's introduction in PTOLEMY (1966), Quondam (1985).

19. See the editions of MÜNSTER's works listed in the bibliography. There are six sixteenth-century Latin editions of the *Cosmographia*, five published in Basel (1550, 1552, 1554, 1559, 1572), one in Antwerp (1584).

20. 'La gente […] v'è villana, rozza, & crudele, & per tal conto al fare battaglie non poco atta' (MÜNSTER (1558), p. 49).

21. 'Connatro Honel, homo nelle arme illustre […] Havardo re d'Inghilterra […] si governo quietamente. Morto costui Thomaso Giraldino Capitanio dell'Hybernia, ricordandosi che suo padre era dagl'Inghlesi stato ucciso, comosse tumulto intorno all'anno 1534, & armò oltre a quaranta milla persone tra pie & cavallo: il qual moto fu di poi con gran pericolo & danno degl'Hyberni, & con certe conditioni quietato' (MÜNSTER (1558), p. 49).

22. 'Templa eorum plerunque in sylvis editisve in locis sita sunt, praesules habentes eruditos, ad canendum docendumque aptos, prasertim in reformatis ecclesijs' (MÜNSTER (1572), p. 43–4).

23. 'Parenti nobili et Catholici' – 'cadendo per colpa della Balia in terra, quella (distendendo le picciole braccia) mostrò abbracciare, quasi dinotando il mirabile studio, che doveva intorno quella fare; si come chiaramente si manifesta nell'opera della sua Universal Fabrica del Mondo' (from 'La vita dell'Eccellente Gio. Lorenzo d'Anania', in ANANIA (1596)).

24. 'L'Isola d'Hirlanda, [che i paesani hor nomano Erin]' (ANANIA (1573), p. 6r [ANANIA (1596), p. 3]); 'questa felice parte' – 'maravigliosa temperie' (ANANIA (1573), pp. 5v–6r); 'maravigliosa bontà' (ANANIA (1596), p. 3); 'per occolto mestiero della divina providenza, dove non essendo cose naturali, gli occhi della ragione vi s'abbagliano, quasi notturno uccello per li raggi del Sole' – 'nefanda bestialità' (ANANIA (1573), pp. 6v–7r; ANANIA (1596), p. 4).

25. 'Nove grotte' (ANANIA (1596), p. 4); 'per humana curiosità' – 'terrore e spavento' – 'mezo selvaggi, rozi, d'animo fiero' – 'nel conversare son del tutto Barbari' (ANANIA (1573), p. 6v; ANANIA (1596), p. 4); 'Hor hanno quattro [~~quattro~~: due] metropolitani, con molti soffraganei [~~con molti soffraganei~~ quel d'Armagh e quel di Cassel], liquali [ivi] sogliono essere come nelle bande Orientali, gran parte Monaci, [a cui portano questi Hirlandesi tanto rispetto, che si può con loro guida caminare per tutto da forastieri, con tutto ch'essi siano molto dediti a latrocinij] e perciò danno grandemente opera a gli studij della Theologia Scolastica,

che gli è cagione, che sian rimasti molti di loro Catholici insin adesso' (ANANIA (1573), pp. 6v–7r [ANANIA (1596), p. 4]).

26. 'Valorosi & essercitati nella militia' (ANANIA (1573), p. 7r; ANANIA (1596), p. 4); 'sono tanto vaghi della libertà, che si contentano più tosto morire che soffrire servitù' (ANANIA (1596), p. 4); 'talvolta' (ANANIA (1573), p. 6v; ANANIA (1596), p. 4).

27. 'Sotto il dominio de gli Inglesi' (ANANIA (1573), p. 7r; ANANIA (1596), p. 4); 'vestimenta di tela gialla, con la portatura alla Cingaresca [...] sono per la conversatione de i forastieri civili & amorevoli e vesteno quasi all'Inglese' (ANANIA (1573), p. 6v; ANANIA (1596), p. 4).

28. On Botero see Descendre (2009), Firpo (1971: 'geopolitical manual', p. 357), Headley (2000).

29. 'Animo nobile' – 'fanciulli quei, che non hanno cognitione di altra parte del mondo, che del luogo, ove sono nati, che terminano il loro pensiero co' confini de' loro poderi, e 'l corso dell'intelletto con le loro ville' – 'propagare à tutto potere la fede' – 'La podestà della Chiesa Romana non hà termine; la giurisdittione della Sedia Apostolica non hà fine. Passa i monti, trapassa l'Oceano, abbraccia l'uno e l'altro emispero, & ha per impresa d'indrizzare tutte le genti à Dio, e per fine di alzar tutta la terra al cielo' – 'havendomi S. Signoria Illustrissima [...] imposto che io le descrivessi lo stato, nel quale si trova hoggi la religione Christiana per il mondo' (BOTERO (1591), dedication to Charles of Lorraine).

30. 'La parte più habitata di quest'isola è l'orientale e la meridionale sino à confini di Mononia, nel quale spatio ci sono due Arcivescovati, Armagnac, che è la metropoli, e Cassel. Sotto Armagnac sono le città e le terre orientali, sotto Cassel le meridionali. Ma la città più importante e più nobile si è Dublin, con guardia d'Inglesi, à i quali soggiaciono le parti più civili e più habitate. Ma le altre che sono verso settentrione e ponente, Ultonia, Connacchia, Mononia, ubidiscono à diversi Signori, assicurati da paludi e laghi e boschi, che sono le loro fosse, e terrapieni e castella, ma più dalla povertà, non vi essendo che guadagnare con esso loro; e sono più sicuri di estate che d'inverno, perché l'acque, che d'inverno agghiacciono, d'estate allagano i campi. E questi più che gli altri si sono mantenuti nella sincerità della fede Catolica contra l'arti e tirannie usate dagl'Inglesi per infettarli dell'empietà di Calvino e di Zvinglio. Gli habitanti che, come scrive Strabone, stimavano cosa laudabile il mangiare i loro genitori morti, hanno ancora adesso dell'agreste e del salvativo assai. Sono nimici della fatica, e da poco. Egli è vero che quei, che habitano alla marina, hanno, per la pratica de' forastieri, dell'amorevole e del civile tanto quanto' (BOTERO (1591), p. 240).

31. 'Di quest'isola si contano cose, per le quali l'Ariosto la chiamò favolosa' – 'chi capita quì, e vi dorme di notte, è travagliato estremamente da gli spiriti maligni. Questo lugo si chiama il Purgatorio di S. Patritio' (Botero (1591), p. 240). 'Diletto' (Botero (1591), dedication). 'Potrà [...] pigliar porto nella famosa Inghilterra, ò nella favolosa Hibernia' (Botero (1618), dedication of Part 1, vol. 2 to Cardinal Pietro Aldobrandino).

CONCLUSION

Mad World
Forever Imagining Ireland

Authors and works discussed in this chapter (in chronological order):

- Giovanni Pontano (1422–1503), *The Ass* [*Asinus*] – also see Chapter Five
- Belisario Acquaviva (1464–1528), *On Hunting and on Falconry* [*De venatione et de aucupio*]
- Niccolò Machiavelli (1469–1527), *The Prince* [*De principatibus*] – also see Chapter One
- Torquato Tasso (1544–1595), *Jerusalem Delivered* [*Gerusalemme liberata*].
- Giovanni Giovane (late sixteenth century), *The Antiquity and Changing Fortune of the Inhabitants of Taranto* [*De antiquitate et varia fortuna tarentinorum*].

To avoid having to surprise the readers is probably the reason why in his epic poem *Jerusalem Delivered* [*Gerusalemme liberata*] Torquato Tasso (1544–1595), arguably Italy's greatest poet and the writer whose words serve as an epigraph for this book ('Whatever pleases the most powerful is right'), decided to coin what is surely the most memorable definition of Ireland, besides Ariosto's *Ibernia fabulosa*, to have emerged from Renaissance Italy: 'furthest Ireland divided from the world' ['la divisa dal mondo ultima Irlanda']. To his readers the expression would not only have evoked Virgil's *ultima Thule*, with its dream of seeing Rome's imperial destiny becoming manifest in the fullness of its glory, as we said in Chapter One, but it would also have added to the verisimilitude of the poem, and verisimilitude, or realism, was one of Tasso's obsessions. The poem is set at the time of the first crusade (1096–99), which for its promoters, as well as for Tasso, was indeed an enterprise aimed at realizing and glorifying (Christian) Rome's imperial destiny, and at that time Ireland – in this case sending troops to assist the Christian forces, in a replication of Ariosto's recruiting of 'hairy beasts' from 'beyond' to help Charlemagne ('Here come the English archers, and men with them who dwell closest to the pole; covered in hair, they have been sent by the woods of furthest Ireland, divided from the world') – would indeed have been, or at least perceived to have been, 'furthest' and 'divided from the world'. We, though, the readers of today, will find it hard not to be surprised.

What is surprising is that, notwithstanding Ireland's apparent location at the end of the world, Tasso, in the very same poem, makes a point of depicting a world where furthest is even further. Using poetic licence and necromancers with the same mastery as Ariosto did, he is able to make the eleventh century foresee a time when 'to a new pole, oh! Columbus, thou shallst so far unfold thy fortunate sails, that fame, with thousand eyes and thousand wings, will hardly stay the course'. What also seems surprising is that Tasso was an avid reader of d'Anania's *Entire Edifice of the World*, which we know he annotated carefully. Admittedly d'Anania's Irish, 'Scholastic robbers' that they are, do appear to stretch the bounds of credulity to the point where his readers are likely to have felt that Ireland was far removed from the rest of the world, but they nevertheless do possess qualities which, to us at least, makes their exile at the hands of Tasso seem unwarranted. It is almost as though all that had passed since the days of Petrarch and the rediscovery of the ancients had not really happened at all. Ireland was still as unworthy of being remembered, at least for governmental purposes, as it had been for Strabo. That, however, is the nature of the language-games we play with writers of the past, especially when history is uncooperative and does not progress in the way we expect. We are bound to be surprised, and to ask the wrong questions.[1]

This is not the first time we have been wrong-footed in our expectations. To some of us it may have seemed extraordinary to find out that Renaissance Italy knew nought of Irish saints and scholars. But when we read what follows, which is taken from the preface to the late sixteenth-century *Antiquity and Changing Fortune of the Inhabitants of Taranto* [*De antiquitate et varia fortuna tarentinorum*] by the little known local historian

Giovanni Giovane, we may be still more perplexed. 'When the inhabitants of Taranto went back to the worship of idols', writes Giovane,

> just as dogs go back to sniff at vomit, one hundred and sixty years after our Lord rose to His Father, during the pontificate of Anicetus, the blessed Cataldo, coming from the island of Hibernia, which is in the ocean, brought them back to their earlier faith. And in order to do this he had been called by God and had renounced his intention of leading a hermit's life in the area of Jerusalem. And if earlier he was a most glorious and most famous saint, because of the miracles he performed in the whole of Italy as well as in Apulia, now he is our protector and our patron saint, by virtue of a confirmation granted by Pope Gregory XIII.

Never mind that Giovane plays tricks with chronology (St Cathal lived a good four centuries later than he says), what is particularly puzzling is that the inhabitants of Taranto, including Giovane himself, appear to have forgotten where Ireland actually was (somewhere 'in the ocean') and not to have cared particularly. Laying claim to Cathal – '*our* protector, *our* patron saint' – was what was of the essence. Cataldo was an *Italian* saint and scholar, and Hibernia was too far to matter.[2]

Not knowing where your patron saint comes from is one thing, but not knowing where the Irish goods you purchased come from is most definitely the sign of a world gone mad – like our own, surely, where a Made in Italy label is just as likely as not to conceal a Bangladeshi provenance. Among the goods exported from Ireland, we may remember, Paolo Giovio lists falcons 'greatly prized for the speed of their flight and destined for the hawking of kings', and Giovio was not alone in thinking highly of Irish falcons. Gerald of Wales, as we saw, also mentioned them, and so did d'Anania, but more importantly, perhaps, they receive a special mention in a work published in Naples in 1519 and subsequently disseminated throughout the rest Europe. The author of the work is Belisario Acquaviva, Duke of Nardò (1464–1528), a member of one of the most powerful baronial families in the Kingdom of Naples (the Fitzgeralds of Naples, so to speak), who together with his elder brother Andrea Matteo, Duke of Atri (1458–1529), the husband of a great-niece of Pope Pius II, was among the most enlightened promoters of humanism in Italy's biggest and most important state; and the purpose of the work was primarily to allow Belisario to share his enlightenment and education with members of his class. It is a four-fold treatise and in it the Duke of Nardò advises his fellow princes on how to educate their children (*On Educating the Sons of Princes* [*De instituendis liberis principum*]), how to safeguard their patrimony (*A Paraphrase on Aristotle's Economics* [*Paraphrasis in Oeconomica Aristotelis*]), how to better their vocation (*On Military Matters and on the Duel* [*De re militari et singulari certamine*]), and not least, since the true sign of a nobleman is that he does not need to work, on how to fill their leisure time (*On Hunting and on Falconry* [*De venatione et de aucupio*]).[3]

Falconry was indeed the sport of princes *par excellence*, as made clear by Giovio, and the acknowledged expert on the matter was Emperor Frederick II (1194–1250), the German Hohenstaufen prince who in 1197 had inherited the Kingdom of Sicily and Naples from the Normans, and who was a passionate falconer himself, devoting much of his own leisure time to the composition of what is still regarded today as an invaluable manual on the matter, the hefty *Art of Hunting with Birds* [*De arte venandi cum avibus*]. No detail of the art is left untouched by Frederick, and he pays particular attention to ensuring that his readers know where to find the very best of falcons, principally gerfalcons. 'Gerfalcons', he therefore writes, 'also nest on an island which is between Norway and Greenland, and in German is called Iceland [*Yslandia*], which in Latin translates as frozen place or region, and these are better than all the other gerfalcons'.[4] Belisario, though, knew otherwise. 'Gerfalcons', he writes in *On Hunting and on Falconry*,

> also nest on an island which, in the vernacular, is called Ireland [*Hirlandia*] and which, compared to others, is such a cold region that no other kinds of falcons can be found there. It is a most infertile land and uninhabitable, because of very high mountains and many rocks and excessive frosts and cold weather. Because of this, it is not possible to find any type of wheat there and the inhabitants feed on nothing else but cold-dried fish. Only one side of the island, which looks towards the east and is washed by the sea, is inhabitable. The other parts, which look towards the west, the north and the south, are impassable and uninhabitable because of the excessive cold. During the months of November and December, when the sun has climbed to the middle of their sky, it sets after only two hours. But during the months of June and July the day there lasts forever. And it is not surprising that, compared to others, that region should be so cold, for the rays of the sun always fall at an angle there. As it crosses the sky the sun circles around their heads, nor do its rays ever fall straight down on the island. The merchants who go to the island import some wheat and

> flour and merchandise of no value, which they are said to exchange for dried fish. It is they who brought gerfalcons to Maximilian, emperor of the Germans.[5]

Belisario mentions many other countries where falcons and gerfalcons can be found, but Ireland is the only one he pauses to describe, in a passage, what is more, which adds little to his argument. He is not arguing that Irish gerfalcons are the best *because* they come from that 'most infertile and uninhabitable' island. He is simply describing Ireland for the sake of doing so, or possibly to impress and satisfy his readers.

Quite why he should have wanted to do so, and do so the way he did, is something of a puzzle. Not only is the *Hirlandia* he imagines so totally different from the one we are familiar with and not only, according to today's experts, have gerfalcons never been known to nest in Ireland – that, however, as stated in the Introduction, is to go 'beyond the limits of our plan' (to quote the words of Pius II) and so we shall not go there – but Belisario's *Hirlandia* is totally different from any of the other Hibernias imagined in the Renaissance, and what is more, Belisario is the only Renaissance or earlier expert on falconry to mention gerfalcons from Ireland. Neither Frederick II, nor 'Maximilian, emperor of the Germans' (i.e. the Holy Roman Emperor Maximilian I: 1459–93–1519), nor any of the other authors writing about falconry do so.[6] Perhaps, having misunderstood 'the merchants who go to the island', Belisario wished to establish his own credentials by calling into question the infallibility of his mentor – *that* 'frozen place', he may intend to signal, is called not *Ys-landia* but *Hir-landia* – and possibly too, as we shall shortly be finding out, to play games with his teacher, Giovanni Pontano. In truth the matter is hardly of the utmost consequence and it may never have puzzled anyone but the most pioneering of readers among the princes of Naples (a class notoriously averse to reading and learning, especially in Latin) – were it not for Machiavelli.

In the *Prince*, first published posthumously in 1532, Machiavelli (1469–1527) had provocatively set out to alter people's perception of the truth. It was his intention, he claimed, 'to write things which are useful to those who understand them' and he therefore proposed 'to pursue the actual truth of things rather than their imaginings'. The 'actual truth' for which he is most famous, of course, it that princes, in order to be good, need to learn how to be bad. 'It has to be understood', he pronounces in the most controversial chapter of the work (Chapter Eighteen),

> that a prince, and especially a new prince, cannot heed all those things by which men are considered good, being often obliged, in order to maintain his state, to act against his word, against kindness, against humanity, against religion. And therefore he must have a mind which is prepared to change in accordance with what is commanded by the winds of fortune and the variations of things and, as I said above, not take leave of good if he can, but know how to enter into evil if he must.[7]

Not everyone, however, agreed with Machiavelli's version of the truth, and many were determined to oppose it. So it was that in 1578 the Basel publisher Pietro Perna (c.1520–82) brought out a volume, containing a number of treatises on the subject of the prince, the aim of which was, precisely, to prove that there were other and better ways for a prince to be good than Machiavelli's; and one of the treatises included in the volume was Belisario Acquaviva's. Therefore any anti-Machiavel worth his salt would sooner or later have found out about that cold, infertile, uninhabitable, mountainous, wheatless, dayless and nightless island called *Hibernia*, prized only for its dried fish and its gerfalcons – just as any Dante fan worth his salt, let us remind ourselves, would sooner or later have been bound to find out about the Fitzgeralds of Florence. Though Ireland was 'furthest' and 'divided from the world', and though there were not that many authors, all told, who had written about it, and certainly not many who had actually been there, Petrarch, in a sense, had been right: Ireland was 'known [...] almost as well as Italy and France'.

The one important question that remains to be answered, therefore, is: what was it about *Ibernia fabulosa* that, to all intents and purposes, made it season-proof and gave writers, whether at the beginning or the end of the period under examination, the licence to imagine it as they wished and as it suited them, that enabled them not to need to know more about it than they wanted to, that convinced them, in most cases, that its fables were its truth? Did it really have to do, as this book has broadly argued, with the specific way humanists constructed knowledge, pursuing the 'imaginings of things', to use Machiavelli's words, rather than the 'actual truth of things'? Was it indeed, as has repeatedly been stressed, that the truth is always the truth *for* rather than the truth *about*, and in particular the truth *for* 'the most powerful'? Or had it to do specifically with Ireland?

To answer the question fully it would of course be necessary to examine the way Italians of the time, and in particular the authors discussed above, viewed other countries of the world, especially 'furthest' countries. Failing that – and going back in time, so as to prove our point that history does not necessarily progress in the way we expect it to and that it is possible to be modern at any time – we could do worse than seek enlightenment from one of the major figures of late fifteenth-century Italy, namely Giovanni Pontano (1422–1503), whom we have already encountered, in Chapter Five, as an authority on historiography, but whom in this case we meet as the only Italian author, as far as we know, to have written not of Italians travelling to Ireland, whether in reality or virtually, but of Irishmen travelling to Italy, and what is more, doing so with scant imagination.

Nowadays, when he is all but forgotten except by specialists, it is difficult to conceive just how famous and important Pontano was in his own day. Originally from Umbria, as a young man he was talent spotted by the king of Naples and in due course rose to become the kingdom's prime minister, playing a leading role in most of the major events of the day. In the work under consideration for instance (a dialogue or playlet entitled *The Ass*), we get to know him as the chief architect of what is hailed as the new *pax romana* – but in fact turns out to be a *pax neapolitana* – that has put an end to the war between the king of Naples, Ferdinand of Aragon (1431–58–94), and his then enemy, Pope Innocent VIII (1484–92), an achievement which sees Pontano compared (no doubt with a strong dose of self-irony) to 'the ruler of heaven and guardian of earth', that is, to God himself.[8]

In earlier years Pontano had acted as tutor to the heir to the throne, the Duke of Calabria, for whose benefit he had authored a work also entitled *The Prince* [*De principe*], and he had taught the sons of other noble families as well, including the Acquaviva brothers, who were to be his star pupils. In addition, for the whole of his life he remained an indefatigable writer, who set himself the challenge of essaying all the literary genres mastered by the ancients. He wrote exclusively in Latin and became famous above all for his love poetry, but he also composed other types of poetry, histories, moral treatises, scientific works, dialogues and much else besides. What he saw as his major and most ambitious work however (his *Iliad*, so to speak: it was the ambition of every self-respecting humanist to write an *Iliad*, or an *Aeneid*) was a poem in epic hexameters, titled *Urania*, which deals with astrology, astronomy and cosmography – or as d'Anania might have said, with the 'entire edifice of the universe' – and in which Ireland figures in the way it does in *Jerusalem Delivered*, though much more menacingly, with an 'army-potent god' who, 'in disorderly commotion, rushes eagerly to war from the arctic cliffs of frenzied Hibernia'.[9] The menace of Ireland, as we will soon find out, was what intrigued Pontano.

Although an extremely busy man, Pontano was also a man who took his *otium*-time very seriously. As one of the speakers in *The Ass* remarks, when he was in Rome, even though he was driven to distraction by the peace negotiations, he still managed to attend to the composition of *Urania*, to visit ancient sites, to enjoy good meals in the company of friends, and to frequent brothels. His greatest contribution to the pursuit of *otium* was to turn the academy which still bears his name today (*Accademia Pontaniana*) into the centre of the kingdom's intellectual life. Now a learned society, the main *raison d'être* of which is the publication of an academic journal, in those days the academy, like the other academies that were emerging in many parts of Italy (not least Florence's Platonic Academy, which, almost at the very time *The Ass* was being composed, stood godfather, as we saw, to Berlinghieri's *Geographia*), was intended as a reincarnation of the academies of ancient times, primarily of Plato's, where men of learning and standing could come together to entertain one another and discuss matters of importance in the company of their peers or betters. The importance of the matters, in truth, varied according to the occasion, and what was discussed was by no means always of the greatest seriousness. It could range from the lofty to the practical, to the gossipy, to the downright smutty. *The Ass*, whose sexual innuendos could cause even today's sophisticates to blush, provides an echo of the *Accademia* at its smuttiest. It deals with serious matters, but it does so in a spirit which the *Pandineschi* and other villa-bound conformists would no doubt have lost no time in gaggling into silence.

Being an account of what went on in the Academy, which in many ways was like a secret society, and being intended, possibly, for a limited audience, which means it is like an insiders' joke, *The Ass* is a language-game that for us is not always easy to follow. Ostensibly it tells of how Pontano, who was driven crazy by the peace negotiations, has gone mad to the point of falling head over heels for the eponymous donkey, which, the innuendos would suggest, stands for a catamite. It is thanks to Pontano that peace was brought about, but the price he now has to pay is to be at war with his own self. He is one of Naples's most distinguished poets, but he cannot notice how appalling the poems are which the ass insists on writing. He has the worries of 'heaven' and of 'earth' on his shoulders, but he has got himself into a frenzy over the demands of his 'sweetipie' [*deliciolus*], wanting to shower him with precious gifts and attend to his every whim. He is devoted to his learned friends, but he has forsaken them in favour of an ingrate and an imbecile. Still, hope is not all lost. When, having kissed his sweetipie on the mouth, the silly ass responds by biting him, Pontano cries out in disgust: 'You ungrateful beast! That's it. I've learnt my lesson. No more of that. So donkeys farewell now! Farewell then, together with Arcadia, oh you asses!'. The madness, though, is not quite over, for Pontano then turns his lecherous attention onto the bride-to-be of his gardener, a pretty girl suggestively nicknamed 'Little Yeastie' [*Fermentilla*], whose 'little field' he insists it is his *droit de seigneur* to be the first to 'plough' [*arare*]. But the gardener also knows how to defend himself, and he too is able to bite back. You dirty old man! – he shouts at Pontano – 'it'd be better for you if you got down on your knees instead and did a bit of praying [*orare*]'.[10]

The dirty old man had shown no more restraint when talking to his gardener about his love of gardening, dwelling on the ins and outs of planting, grafting and irrigating, expressing scant appreciation of the gardener's performance, telling him not to attach so much importance to the positions of the moon, least of all to its 'copulations' [*coitus lunae*], and advising him instead to look beyond and learn to penetrate the mysteries of the 'sign-bearing heaven' (i.e. the zodiac). Yet what had been the cause of all this madness, it turns out in the end, was not so much the brokering of peace, or the unrequited love of a silly ass, or the pulling power of Little Yeastie, or the less than satisfactory performance of a gardener, as it was the dizzying contemplation of far more lofty matters. When at last Pontano is joined by his worried friends, who have come to find out for themselves whether he is indeed as mad as one of them had made him out to be, the gardener rushes out to warn them that the old man has for sure lost his mind and that he is wandering about with his head in the clouds and the heavens in his hands. 'Thank God you're here!' the gardener says to them,

> my master has gone completely bonkers! He's spent half the day turning the heavens around in his mind, and now he's turning them around in his hands. If I hadn't been there to save him, he would have been killed by the bees that threatened to attack him when he accidentally knocked over their hive while walking around with his head in the clouds. And to cap it all, he's now after my wife!

But as Pontano too comes forward to greet his friends (presumably holding a celestial globe, to cover his discomfiture), he is quite his old self again and fully *compos mentis*. He cannot hide his delight at seeing them, having spent the last few days, he tells them, contemplating in solitude upon 'heavenly matters'. Solitary reflection, he concludes, is pointless if it is not followed by communication with men of learning, for communication with men who are experts in the same matters as oneself is, for sure, the sweetest fruit of contemplation.[11] So what *The Ass* is all about, it would seem, is the withdrawal symptoms one has to suffer when unable to play language-games with one's peers and betters. Yet if that is so, what could be the role of the Irish – or better perhaps, to quote Berlinghieri, the 'errant Irish' – who appear in the first act of the playlet?

As the playlet opens, at a tavern outside Naples, an inebriated taverner and an equally inebriated traveller are celebrating the return of peace, which they greet with undisguised relief. 'Peace in Rome has been made they say', are the first words of the work. The return of peace means that life can go back to normal, and the first sign of normality is the news that pimps and prostitutes are on their way from Sicily to enliven the festivities. And also that the Irish are in town.[12] But is it quite sure they are Irish? 'Are you certain they are Irish [*Hiberni*]' – the taverner enquires of the traveller, as they watch the group of wayfarers coming towards the inn – and not 'British' [*Britanni*], or even '*Über*-Brits' [*Britannissimi*]? Absolutely – the traveller replies, none the better for drink – look at how 'they bear *fast*idiousness in their nose! They must surely be after that *Fast*inianum wine of yours!'

So get the drinks ready boys! – the taverner answers back, barely able to control his excitement, or his geography – the money will be rolling in! Make room for these travellers from 'furthest Scotland', so that they may 'imbibe in the British manner'. With that, the two revellers get down to some really serious drinking and allow their imaginations to run riot. The taverner will fancy that he is flying across Europe towards Britain and setting up shop in the middle of the ocean, to become immensely rich, while the traveller, seeing double, or even triple, will convince himself he has rediscovered Plato's lost Atlantis or, as he also calls it, the Island of the Blessed.[13]

In the process we find out that the Irish too are on a journey of discovery, though in their case it is an inverted one, so to speak, in search of another type of blessedness. They have come to Italy, the traveller informs us, out of 'stubbornness' [*pervicacia*], for unless they could convince themselves of it with their own eyes, they refused to believe that the Holy Father was a father in every sense of the word. For their good luck, their eyes did not let them down. What they discovered was not only that Innocent VIII did indeed have children, but that at the very moment they arrived in Rome he was presiding from the throne of St Peter at the wedding of a son of his, amid much merry-making, led by one of his own daughters, even as another daughter unto him was born. Is that not sufficient reason to believe? – the traveller exclaims. 'What better proof of the truth of Christianity' could the Irish have asked for? How so? – wonders the puzzled taverner. 'Well – is the answer – if grandchildren are born to God, does it not follow that Christ himself *must* have come forth from the womb of a woman?'[14]

Perhaps that is how the Irish felt as well, for their stubbornness is unaffected and they still have faith enough to continue with their pilgrimage. After the drunken interlude in Naples, we are informed, they plan to visit St Nicholas of Bari, the resting place of the original Santa Claus and one of the other great pilgrims' destinations of the time, then St Leonard of Apulia, another pilgrims' station on the way to Jerusalem, and finally Jerusalem itself, having in the meantime bought a boat, no less. Or maybe the Irish had premonitions of what was about to pass and felt they had to put as great a distance as they could between themselves and the fathering Holy Father. When, the following year, Lambert Simnel (c.1475–1535), the false pretender to the English throne, went to Ireland for assistance, the

Irish, or some of them at least, responded to his call with little hesitation, to the point of crowning him King of England in Dublin's Christ Church Cathedral, and this, according to a bull Innocent VIII was to issue on the matter, was not only a 'crime of lese-majesty' but an 'affront to papal dignity' worthy of the harshest of punishments. And in a later bull the less than innocent Innocent then threatened to excommunicate any Irish person, and especially any Irish cleric, who dared once again to rebel against the king of England. 'We extend this warning, order and prohibition', he proclaimed (as though addressing *Bergamaschi*, we may be tempted to add)

> to persons ecclesiastic and to the religious, of whatever order, whether exempt or non exempt, that are established in the said kingdom of Hibernia, and in other lands of the said king [...] and we warn them, order them, and prohibit them from provoking and fomenting new rebellions of the sort, or from encouraging, assisting and maintaining rebellions that are already taking place, or from inciting anyone to take part in such rebellions.[15]

In truth we will never know for definite what drives Pontano's Irish on, because they never speak a word, being able to express themselves neither in Italian – as the traveller is keen to stress – nor, presumably, in that perfect Latin which had allowed Guerrino il Meschino to find his way in Ireland without a hitch and for which Giovio, and even Virgili, were to give the Irish every credit.[16]

Perhaps the Irish of *The Ass* represent the *sous-entendu*, the unspoken, or even the unspeakable, in a world of language-games where everybody is continuously 'on the go' [*in itinere*] – *iter* (journey, travel) and *via* (way, road) are the dialogue's dominant metaphors –, where language is equally as restless, and where nothing means what it ought to mean, with names and words running amok, and signifiers 'copulating' randomly with signifieds. In that world, an ass is a human, *pax romana* is *pax napolitana*, being Innocent means proving the truth of Christianity with the power of your groin, ploughing is a metonymy for praying, heterosexual love is bisexual, since it brings together man and woman, and the Irish are British, or even *Über*-Brits who hail from furthest Scotland. Is it any wonder that when the taverner promises to reward his customers with a real meal, the traveller replies: 'What do you mean, that "dinner should be dinner"? I do not understand'?[17] It is a world, in short, for which there is no proper language.

The reason that language fails that world is that – in an amazing forecast of things to come, as though Pontano had carefully studied Fra Mauro's map and grasped how earth-shattering the great discoveries would be – the world itself is threatening to change beyond recognition. On the face of it, it is a cosy world, safely ensconced within *mare nostrum*. At its heart lies Rome, to the east it is bounded by 'happy Arabia' [*Arabia felix*], to the west by the Island of the Blessed, to the north by *ultima Scotia*, and to the south by the meretricious Spain of pimps and prostitutes whose advent heralds the return of the good old days. And it is a world whose cosiness is safeguarded by the dream of Arcadia, or in other words (we must presume) by the spurious charms of an earthly paradise located full square within the confines of the Mediterranean, a paradise which Pontano's friend Iacopo Sannazaro (1457–1530) – who is present in *The Ass* under the nickname *Actius* – was at that very time making into the subject of what was to become his hugely popular pastoral novel *Arcadia*. That, however, was the problem: Arcadia was but a dream (or possibly a nightmare), and the cosiness of the world was meretricious – which is why (and that is the only way of making sense of what otherwise seems like an illogical utterance) Pontano had to take leave of it, even as he did of his beloved *Deliciolus*: 'Farewell then, together with Arcadia, oh you asses!'[18]

As it happens, what makes the dream of asses and Arcadia so impossible is the advent of the Irish. With *ultima Scotia* now threatening the 'innocence' of Rome, and with what was 'furthest' now becoming nearest, *mare nostrum* is no longer *nostrum*. Indeed, it is no longer *mare* by antonomasia, since it now has to compete with the ocean whence the stubborn Irish hail and whither the inebriated taverner has fantasies of taking flight to become wealthy; and Rome, as is explicitly pointed out, is now 'under siege'.[19] It is to Naples that all roads will henceforth lead – and Pius II for one, whose *De Europa* concludes with Naples, would certainly have agreed with that. What is more, new (lost) lands were being discovered in the wide yonder and blessedness – as the *Pandineschi* were so desperately to fear – looked likely to be found elsewhere. A new madness was set to take over from the old, and though the Self had regained composure – had cared sufficiently for itself to regain composure – a new Other was threatening to engulf it. It would be *pervicacia* to deny it. There would be no telling who was outside the cage laughing, and who inside being laughed at.

Faced with a situation like this, the strategy of the Countess of Pandino would be to giggle and cast her imperious eyes at it, so as to deny it the power of encroaching upon the safety of her own earthly paradise, which as far as she was concerned contained as much of the earth as was needed. The strategy of Pontano's 'errant Irish' was *pervicacia* (an attribute, incidentally, which in *The Ass* only they and the eponymous donkey are said to share). It was to carry on regardless of what your eyes told you, pretending that they did not tell you what you had hoped they would not, or in other words, continuing to be pilgrims even though sanctity had been sucked out of your pilgrimage at the very heart of Christendom. It is a strategy Pontano's friends, and possibly Pontano himself, would have approved of, since one's eyesight could often let one down. Indeed, is it not just as well – one of the friends suggests to the others – that they did not see what the old man was up to with his donkey? The question, however, is far less innocent, far more subversive than may appear at first. It concerns far more than silly asses and 'errant' Irishmen. In siding with the Irish Pontano and his friends were in effect taking sides against eye-witness travellers, for that is what the Irish are. What *is* the point of seeing things with your own eyes? – is what they appear to want to ask us to ask ourselves, between the licentious and meretricious lines of the playlet. Is it not the case that seeing things with your own eyes may not make you see what you *want* to see? And that it will certainly not make you see what you do *not* want to see? Above all, is it not the case that seeing things with your own eyes will not affect the reality of those things? It is not *because* the Irish catch sight of him that Pope Innocent is an Unholy Father. He is what he is, regardless. The world *is*, whether we see it or not, and no amount of seeing will make it not be. No amount of seeing it will make us actually see it.

So if seeing is not the solution, what is? For the down-to-earth gardener the best is to trust the 'copulations of the moon', that is, the tried and tested lunacy of the herebelow, or in other words what we might call folklore. The traveller and the taverner are of the same opinion. For all that their hubris carries them into uncharted waters, the world they prefer, just as it was to be for the *Pandineschi*, is the cosy world they know, where pimps and prostitutes make their way from Sicily and the Irish get drunk (incidentally, we are only ever *told* that the Irish get drunk, by the dipsomaniac traveller and innkeeper; we never actually see them in that state, so it is unlikely they represent the stereotypical Irish drunkards one can apparently find in other literatures at the time).[20] Yet what good can possibly come of lunacy, of inebriation? It is simply another form of *pervicacia* or of madness. But not even 'contemplation', as it turns out, is to be trusted, for the heavenly signs – the semiotics of the divine – are in truth no easier to penetrate than the human ones. Pontano nearly killed himself as he tried to do so. Thus, when all is said and done, all that seems to work, all that can deliver us from madness, is communication with one's friends, communication with like-minded people, communication with men of learning. Truth, as we have said repeatedly, is a moving target, a rhetorical construct. More often than not it is truth *for*. So all that can work are language-games.[21]

Pontano, there is no doubt, was uncommonly prescient, not to say modern. His foresight was a match for anyone's sight, not only with regard to the world that was changing before the very eyes of Italy and Europe (or better, perhaps, before their very imagination, their imperious imagination), but also, apparently, with regard to Ireland. He had understood, it would seem, that Ireland was at once menacing and tame, that it could and would be imagined or constructed as an Other – a silent, voiceless, and mostly faceless Other – which was nevertheless part of the European Self, an Otherness where the Self could recognize itself by populating it with its own fantasies, which allowed it to peep through the gilded bars of its caged-in selfhood without feeling too threatened, and fully confident that It, not They, was on the outside laughing. Ireland was the Otherness where fables were fables but also the truth, where one's innocence, one's strategies of innocence, could be upheld even as they were exposed. For sure, it was different things to different people. It was Irelands rather than Ireland. For some it was *ultima*, for others all too close. For some a place from which to hurry on and to forget, for others a go-there kind of place, that was fully worth remembering. An inhuman place, inhabited by backwoodsmen and hairy beasts, or an earthly paradise, inhabited by sweet and happy people; a wilderness of wooded mountains, unfit for human habitation, or a sylvan never-never land, for those who needed *libertas* and *otium* and were afraid of lightening and thunder; an island where licit and illicit were the same, or the island of the blessed. Where all were vegetarians, carnivores or cannibals; where St Patrick's Purgatory could save you from yourself and help you find your parents, or where it threatened to engulf you and was too frightening to look at; where kinglets were unfit to rule and acted as *imperium*'s foil, or where they gloried in the name of justice and stood firm in their opposition to *imperium*; where wild geese fled to Scotland, to the Orkneys, to Spain and to Venice. It was of this world, out of this world, and between worlds. It was fully known, and totally unknown. It mattered least, and it mattered most. It was *Hibernia, Ierne, Iwernia, Iuuverna, Iberia, Ibernia, Irlanda, Hirlandia, Irlond* and *Scotia*. It was mad, it was divided, it was amazing. And always it was *Ibernia fabulosa*. Or not.

In short, it was what one's imperious eye-sight, to protect itself, made it out to be. One could look at it through less myopic lenses, as advised by Fra Mauro, but on the whole what worked best were the Rome-tinted spectacles prescribed by Pope Pius II. They allowed one to see only what one needed to see, and they were at their most effective when there was no need to know or the need not to know. There was no point in having twenty-twenty vision to 'see the shape of the world' and 'the habitable face of the earth'. What did the trick – fabulously – was to have twenty-twenty imagination. *And* the ability to feed the hands that could bite. The last laugh is Ariosto's. And the Countess of Pandino's.

Notes

1. 'Tu spiegherai, Colombo, a un novo polo/lontane sì le fortunate antenne,/ch'a pena seguirà con gli occhi il volo/la fama c'ha mille occhi e mille penne' – 'Sono gl'Inglesi sagittari, ed hanno/gente con lor ch'è più vicina al polo:/questi da l'alte selve irsuti manda/la divisa dal mondo ultima Irlanda' (TASSO (1967), vol. 2, p. 485: 15.32; vol. 1, p. 19: 1.44). For an introduction to Tasso see Brand (1965). On Tasso's reading of d'Anania see Basile (1984), p. 335. For more on Tasso's sense of geography and his struggle with verisimilitude see Gori (1990).
2. 'Quando [...] i Tarantini ritornarono al culto degli idoli, come il cane torna al vomito, centosessanta anni dopo l'ascesa del Signore al Padre, durante il pontificato di Aniceto Siro, il beato Cataldo, provenendo fin dall'isola Ibernia che

si trova nell'oceano, li riportò alla fede precedente. Per far questo era stato chiamato da Dio e per questo aveva trascurato il proposito della vita eremitica nelle parti di Gerusalemme. E se prima era santo gloriosissimo e famosissimo per i miracoli compiuti in tutta l'Italia nonché nella Iapigia, ora è il nostro patrono e santo tutelare in virtù di una conferma voluta dal pontefice Gregorio XIII' (translated into Italian by and quoted in Tateo (2002), p. 191).

3. On the Acquavivas see Defilippis (1993).
4. 'Nidificant autem girofalci [...] in quadam insula, quae est inter Norvegiam & Gallandiam, & vocatur Teutonice Yslandia, & Latine interpretatur contracta, seu regio glaciei [...] & isti sunt meliores & nobiliores illis qui nidifcant in remotis a maritima' (FREDERICK II (1596), p. 152).
5. 'Nidificant [hierifalcones] etiam in insula, quam Hirlandiam vulgo nominant adeoque prae caeteris frigidissima est regio, ut aliorum ibi generum accipitres non inveniantur. Infertilissima quidem altissimisque montibus & saxis plurimis geluque nimio ac frigore inhabitabilis. Quam ob rem nullius ibi frumenti genus reperiri contingit, nec nisi siccis frigore piscibus homines vivunt. Insulae latus unum ad orientem spectans, quod mari abluitir, tantummodo habitabile est. Alia, quae ad occidentem septemtrionem et meridiem spectant, invia frigoribusque nimiis inhabitata. Novembri mense decembrique, postquam ad illius coeli medium sol ascenderit, duabus tantum horis sol totus occidit, sed iunio & quintili mense dies ibi continuus. Nec mirum quod frigidissima prae caeteris illa sit regio, qum solares ibi radii semper obliqui feriant solque illa incedens supra eorum caput circumferatur, nec solis umquam rectis radiis insula feriatur. Mercatores, qui insulam adeunt, frumenti farinaeve parum ac vilissimas ibi merces important, quae siccis piscibus commutari dicuntur. Hi quidem hierifalcones accipitres ad Maximilianum Germanorum imperatorem deferunt' (ACQUAVIVA (1519), *De aucupio*, p. XVIr).
6. For more on Belisario and authors of falconry see Haywood (1999), pp. 380–1.
7. 'Sendo l'intento mio scrivere cosa utile a chi la intende, mi è parso più conveniente andare drieto alla verità effettuale della cosa, che alla imaginazione di essa' – 'E hassi ad intendere questo: che uno principe, e massime uno principe nuovo, non può osservare tutte quelle cose per le quali gli uomini sono tenuti buoni, sendo spesso necessitato, per mantenere lo stato, operare contro alla fede, contro alla carità, contro alla umanità, contro alla religione. E però bisogna che egli abbi uno animo disposto a volgersi secondo ch'e venti della fortuna e le variazioni delle cose li comandano, e, come di sopra dissi, non partirsi dal bene, potendo, ma sapere intrare nel male, necessitato' (*The Prince*, Chapters Fifteen and Eighteen: MACHIAVELLI (1976), pp. 94, 100).
8. 'coeli rector [...] terrae tutor' (PONTANO (1943b), p. 292).

9. 'Deus armipotens [...] incomposito confestim in bella tumultu/currit ab arctois scopulis furialis Iberne' (PONTANO (1902), pp. 152–4: 5.137, 176–7).
10. 'Apage te, bestiam nequissimam! Hoc, hoc illud est, tarde illud didici senex improvidus [...] O asini, valete iam, valete posthac ipsa cum Arcadia, asini!' – 'PONT. Hoc ipsum est, mi Caselio: arare ego tecum in luce una cum uxorcula agellum velim. FAS. Orare profecto ad genua quoque provolutus potes' (PONTANO (1943b), pp. 305–6).
11. 'coitus lunae' – 'signiferi orbis cognitio' – 'Adite ad illum, liberabitis eum alia etiam molestia. Diem hunc pene dimidiatum transegit in volutando coelo, quod in manibus adhuc etiam versat; cui mea quidem sententia cerebrum identidem volutatur, et, ni ego accurrissem, amiculo iniecto, dum coelum suspectat et alvearia non videt, in ea illatus, apiculae illum confecissent [...] suboluit enim illi uxorculam mihi ducendam, nihil enim est eo salacius an nescio quomodo in maritas foemellas magis accenditur' – 'contemplationis ipsius communicatio cum studiosis rerum earundem viris is profecto fructus est etiam suavissimus' (PONTANO (1943b), pp. 300–1, 308).
12. 'Pacem Romae factam esse aiunt' (PONTANO (1943b), p. 292).
13. 'Sed amabo, Hibernine sunt hi?' – 'Hibernici generis homines, fastum in naso qui gerunt' – 'An forte verendum est ne hodie Fastinianum hoc tuum nobis exhibeat negocium?' – 'Scotia nuper ab ultima Romam devecti' – 'Britannice ut discumbant' – 'beatorum insula' (PONTANO (1943b), pp. 292–5).
14. '... mirificam, mihi crede, Christianae religionis comprobationem' – 'Si enim Deo nascuntur nepotuli, nunquid non necesse est Christum ipsum mulieris utero prodiisse?' (PONTANO (1943b), p. 294: emphasis added).
15. 'Crimen leasae majestatis' – 'dignitatis pontificalis opprobrium' – '... ac easdem monitionem, requisitionem, et inhibitionem ad personas ecclesiasticas, etiam cujusvis ordinis religiosas, exemptas et non exemptas, in praefatis regno Hiberniae, et aliis dominiis ipsius regis constitutas [...] extendimus et [...] monemus, requirimus, et inhibemus eisdem, ne novos tumultus hujuscemodi suscitare, movere, seu jam motos fovere, nutrire, et manutenere, seu quempiam ad illos incitare' (*Innocentii VIII bulla contra hibernicos praelatos, qui Lambertum Symnell praetensum de jure de facto in regem coronarunt* & *Innocentii VIII bulla contra rebelles domini regis*, in D. Wilkins (1964), pp. 623–4).
16. 'They understand nothing in Italian' ['Italice nihil intelligunt'] (PONTANO (1943b), p. 295).
17. 'Quid hoc verbi est "coena ut coena sit"? Non intelligo' (PONTANO (1943b), p. 294).
18. *Arcadia* recounts the amorous woes of a love exile from Naples who has sought refuge in the eponymous region of Greece. For a readable, though not very

penetrating, introduction to Sannazaro see Kidwell (1993). See too Haywood (2009c).

19. 'Obsessa Roma' (Pontano (1943b), p. 290).
20. See Hadfield (2004), p. 45.
21. A more detailed analysis of Pontano's *Asinus* can be found in Haywood (2003b). And a more recent edition of Pontano's *Asinus*, with bibliography, can be found in Tateo and Canfora (2004).

Bibliography

AUTHORS spelt with small capitals in the footnotes to the various chapters (= authors of primary texts) are listed in Part One below. Authors spelt in lower case (= authors of secondary texts) are listed in Part Two. Only for the authors of primary texts are first names given in full.

The names of the authors are followed (in brackets) by the date of publication or republication of their works. In the case of more than one work by the same author being listed, it is that date which serves, in the footnotes, to identify the specific work that is being quoted. Where more than one work was published in the same year, the date of publication is followed by a small case letter, to indicate sequence. When the date of publication is uncertain, it is followed or substituted by a question mark. In the case of manuscripts, instead of the date, the author's name is followed by (MS).

Where authors (or works) are also known by another name, that other name is given in square brackets after their original name.

Place of publication is given in the original language, e.g. Roma, not Rome. In the case of early printed books and manuscripts I also specify the copy/copies I consulted, since it is not uncommon for copies of the same book to differ from one another.

Part One: Primary Texts

(I list only those editions or manuscripts which I consulted; circumstances did not always allow me to consult the most appropriate or the best.)

ACQUAVIVA, Belisario (1519). *De instituendis liberis principum. Paraphrasis in Economica Aristotelis. De venatione et de aucupio. De re militari et singulari certamine*. Napoli: Pasquet de Sallo. Copy consulted: Firenze, Biblioteca Nazionale, Magl. 1.K.5.263.

ANANIA, Giovanni Lorenzo d' (1573). *La universal fabrica del mondo*. Napoli: Cachij dell'Aquila. Copy consulted: Firenze, Biblioteca Nazionale, Magl. 3.2.443.

ANANIA, Giovanni Lorenzo d' (1596). *L'universale fabrica del mondo*. Venezia: Muschio. Copy consulted: Roma, Biblioteca Nazionale, 6.13.L.18.

ARAGONA, Tullia d' (1560?). *Il Meschino, altramente detto il Guerrino, fatto in ottava rima dalla Sig. Tullia d'Aragona*. Venezia (?): Battista & Sessa (?). Copy consulted: British Library, 83.d.14.

ARAGONA, Tullia d' (2006). E. Pallitto (ed. & trans.), *Sweet Fire. Tullia d'Aragona's Poetry of Dialogue and Selected Prose*. New York: George Braziller.

ARIOSTO, Ludovico (1959). *Orlando Furioso*, ed. N. Zingarelli. Milano: Hoepli.

BANDELLO, Matteo (1928). 'Gandino bergamasco'. In his *Le novelle*, vol. 2 (part 1, *novella* 34), ed. G. Brognoligo, pp. 19–46. Bari: Laterza.

BANDINI, Domenico (MS). Description of Ireland. In his *Fons memorabilium universi*. Vatican Library, MS Pal. Lat. 923, ff. 45v–46r. Also consulted: Vatican Library, MS Reg. Lat. 1140, ff. 268v–269r.

BARBERINO, Andrea da [Andrea Mangiabotti] (1534). *Guerino prenominato Meschino*. Venezia: Bindono.

BARBERINO, Andrea da [Andrea Mangiabotti] (1969). *I reali di Francia*. Milano: Betti.

BARBERINO, Andrea da [Andrea Mangiabotti] (2005). *Il Guerrin Meschino*, ed. M. Cursietti. Roma – Padova: Antenore.

BERLINGHIERI, Francesco (1482a). Dedicatory letter of copy of *Geographia* presented to Sultan Bayezid. Reproduced in Haywood (2009a), p. 106.

BERLINGHIERI, Francesco (1482b). Dedicatory letter of copy of *Geographia* presented to Prince Djem. Reproduced in Babinger (1963), pp. 346–7.

BERLINGHIERI, Francesco (1966). *Geographia*, ed. R. A. Skelton. Amsterdam: Theatrum Orbis Terrarum. Anastatic reprint of the original, Firenze, Niccolò Tedesco, 1482, edition, giving neither page nor line numbers.

BERNARD, Saint (1963). *Vita Sancti Malachiae*. In his *Opera*, vol. 3: *Tractatus et opuscula*, ed. J. Leclerq and H. M. Rochais, pp. 295–378. Roma: Editiones Cistercienses.

BOCCACCIO, Giovanni (1510?). *Opera dell'huomo dotto et famoso Giovan Boccaccio da Certaldo, dalla lingua latina nel thosco idioma per Meser Nicolo Liburnio novamente trallatata, dove per ordine d'alphabeto si tratta diffusamente delli monti, selve, boschi, fonti, laghi, fiumi, stagni, paludi, golfi e mari*. Venezia (?): [s.n.]. Copy consulted: British Library, 304.d.8.

BOCCACCIO, Giovanni (1998). *De montibus, silvis, fontibus, lacubus, fluminibus, stagnis seu paludibus et de diversis nominibus maris*, ed. M. Pastore Stocchi. In his *Tutte le opere*, vol. 7–8, ed. V. Branca, pp. 1815–2122. Milano: Mondadori.

BOIARDO, Matteo Maria (1936). *Orlando Innamorato*. In his *Tutte le opere*, ed. A. Zottoli. Milano: Mondadori.

BORDONE, Benedetto (1965). *Libro di Benedetto Bordone nel quale si ragiona di tutte le isole del mondo*, ed. R. A. Skelton. Amsterdam: Theatrum Orbis Terrarum. Anastatic reprint of Venezia, Zoppino, 1528 edition.

BORDONE, Benedetto (1982). *Isolario*, ed. M. Donattini. Modena: Edizioni Aldine.

BOTERO, Giovanni (1591). *Delle relationi universali*. Roma: Ferrari. Copy consulted: Firenze, Biblioteca Nazionale, RIN.B.426.

BOTERO, Giovanni (1618). *Le relationi universali*. Venezia: Vecchi. Copy consulted: Trinity College Dublin, Fag.K.24.38.

BRUNI, Leonardo [Leonardo Aretino] (1744). *Eloquentissimi viri Leonardi Aretini oratio in funere Nannis Stroze equitis florentini*. In Stephani Baluzii Tutelensis *Miscellanea*, pp. 2–7. Lucca: Vincentius Junctinius.

BRUNI, Leonardo [Leonardo Aretino] (1987). G. Griffiths, J. Hankins and D. Thompson (ed. & trans.), *The Humanism of Leonardo Bruni: Selected Texts*. Binghamton, NY: Center for Medieval and Renaissance Studies, State University of New York – Renaissance Society of America.

CAESAR, Giulius [Gaius Iulius Caesar] (1970). *The Gallic War*, ed. & trans. H. J. Edwards. London – Cambridge, Mass.: Heinemann – Harvard University Press.

CHECCO ROSSI DA MELETO (1984). *Epistola ad magnificum dominum Malatestam Ungarum de Malatestis, in qua plurimum commendatur, et maxime de itinere ad Puteum Sancti Patricii*. Reproduced in Rossi (1984), pp. 16–19.

CHIERICATI, Francesco (1873). Letter to Isabella d'Este about his visit to St Patrick's Purgatory. Reproduced in Morsolin (1873), pp. 87–92.

CHIERICATI, Francesco (1987). M. Purcell, 'St Patrick's Purgatory. Francesco Chiericati's Letter to Isabella d'Este', *Seanchas Ardmhacha* 12 (2), pp. 1–10. (N.B. This translation of Chiericati's letter is not always entirely accurate; in the text I have given my own.).

CICERO, Marcus Tullius (1967). *De oratore*, vol. 1, ed. & trans. E. W. Sutton and H. Rackham. London – Cambridge, Mass.: Heinemann – Harvard University Press.

DANTE ALIGHIERI (2001). *La divina commedia*, 3 vols, ed. U. Bosco and G. Reggio. Firenze: Le Monnier.

DANTE/LANA (1478). *Comedia di Dante degli Allagherii col commento di Jacopo della Lana bolognese*. Milano: Piemontesi. Copy consulted: Roma, Biblioteca Corsiniana, 51.G.5.

DANTE/LANDINO (2001). *Comento sopra la Comedia di Cristoforo Landino*, 4 vols, ed. P. Procaccioli. Roma: Salerno.

FREDERICK II (1596). *Reliqua librorum Friderici II imperatoris de arte venandi cum avibus, cum Manfredi regis additionibus*. Augusta [Aachen]: Praetoris.

GERALD OF WALES [Giraldus Cambrensis] (1964). *Topographia Hibernica et Expugnatio Hibernica*, ed. J. F. Dimock. Wiesbaden: Kraus Reprint.

GERALD OF WALES [Giraldus Cambrensis] (1982). *The History and Topography of Ireland*, trans. J. J. O'Meara. London: Penguin. (N.B. In places, I have adapted O'Meara's translation; and there is not always full congruence between O'Meara's edition and Dimock's, since they are based on different manuscripts of the work).

GIOVIO, Paolo (1956). *Lettere*, vol. 1 (1514–1544), ed. G. G. Ferrero. Vol. 1 of Pauli Iovii *Opera*. Roma: Istituto Poligrafico e Zecca dello Stato.

GIOVIO, Paolo (1984). *Descriptio Britanniae, Scotiae, Hyberniae et Orchadum*. In Pauli Iovii *Opera*, vol. 9: *Dialogi et descriptiones*, ed. E. Travi and M. Penco, pp. 89–125. Roma: Istituto Poligrafico e Zecca dello Stato.

HIGDEN, Ranulf (1865). *Polychronicon*, with English translation of John Trevisa, ed. C. Babington. London: Longman Green Longman Roberts & Green (Rerum Britannicarum Medii Aevi Scriptores 41).

HORACE [Quintus Horatius Flaccus] (1999). *Ars poetica*. In his *Satires, Epistles and Ars Poetica*, ed. & trans. H. R. Fairclough, pp. 442–89. London – Cambridge, Mass.: Heinemann – Harvard University Press.

ISIDORE OF SEVILLE (1966). *Etymologiarum sive originum libri XX*, vol. 2, ed. W. M. Lindsay. Oxford: Clarendon.

LANDINO, Cristoforo (2008). *Poems*, trans. M. P. Chatfield. Cambridge, Mass. – London: Harvard University Press.

MACHIAVELLI, Niccolò (1976). *Opere*, ed. E. Raimondi. Milano: Mursia.

MANNINI, Antonio (1886). Letter to his brother about his visit to St Patrick's Purgatory. Reproduced in Frati (1886), pp. 154–62.

MANNINI, Antonio (1918). 'The Descent of Antonio Mannini'. Chapter 6 of Seymour (1918), pp. 53–61.

MARTIRE, Pietro [Peter Martyr] (1985). 'Minuta descrizione dell'isola Spagnuola'. From *Sommario dell'istoria dell'Indie occidentali cavato dalli libri scritti dal signor don Pietro Martire milanese*. In G. B. Ramusio, *Navigazioni e viaggi*, vol. 5, ed. M. Milanesi, pp. 188–96. Torino: Einaudi.

MASUCCIO SALERNITANO (1979). *Il novellino*. Bari: Laterza.

MAURO Fra (2006). P. Falchetta, *Fra Mauro's World Map*. Turnhout: Brepols.

MEDICI, Lorenzo de' (1992). *Selected Writings*, ed. C. Salvadori. Dublin: UCD Foundation for Italian Studies.

MELA, Pomponius (1988). *Chorographie*, ed. & trans. A. Silberman. Paris: Les Belles Lettres.

MÜNSTER, Sebastian (1544). *Cosmographia. Beschreibung aller Lender*. Basel: Henricpetrinus. First German edition. Copy consulted: British Library, Maps C.8.a.14.

MÜNSTER, Sebastian (1550). *Cosmographiae universalis libri VI*. Basel: Henricpetrinus. First Latin edition. Copy consulted: British Library, 516.i.14.

MÜNSTER, Sebastian (1558). *Sei libri della Cosmographia universale*. Basel: Henricpetrinus. First Italian edition. Copy consulted: Genève, Bibliothèque Publique et Universitaire, VII.1a.MUEN.

MÜNSTER, Sebastian (1572). *Cosmographiae universalis libri VI*. Basel: Henricpetrinus. Last Basel Latin edition. Copy consulted: Trinity College Dublin, HH.e.4.

MÜNSTER, Sebastian (1575a). *Sei libri della Cosmographia universale*. Köln: Byrckmann. Second Italian edition. Copy consulted: British Library, 566.i.1.

MÜNSTER, Sebastian (1575b?). *Sei libri della Cosmographia universale*. Venezia: Thomasini. Third Italian edition.

MÜNSTER, Sebastian (1965). *Mappa Europae*. Wiesbaden: Guido Pressler. Anastatic reprint of Frankfurt am Main, Egenolff, 1536 edition.

PETRARCA, Francesco [Petrarch] (MS). Manuscript of Gerald of Wales *De mirabilibus Hybernie* [*Topographia Hibernica*] owned and annotated by Petrarch. Paris: Bibliothèque Nationale, MS Lat. 4846.

PETRARCA, Francesco [Petrarch] (1950). *Invective contra medicum*, with Italian translation by Domenico SILVESTRI, ed. G. Ricci. Roma: Edizioni di Storia e Letteratura.

PETRARCA, Francesco [Petrarch] (1968). *Le familiari*, vols. 1–3, ed. V. Rossi. Firenze: Sansoni.

PETRARCA, Francesco [Petrarch] (1972). *Invettive contro un medico*, ed. M. Schiavone. Milano: Marzorati.

PETRARCA, Francesco [Petrarch] (1977a). *De vita solitaria*, ed. G. Martellotti. Torino: Einaudi.

PETRARCA, Francesco [Petrarch] (1977b). *Secretum*, ed. E. Carrara. Torino: Einaudi.

PETRARCA, Francesco [Petrarch] (1996). *Triumphi*, ed. V. Pacca. In his *Trionfi, rime estravaganti, codice degli abbozzi*, ed. V. Pacca and L. Paolino, pp. 3–626. Milano: Mondadori.

PIUS II [Eneas Silvius Piccolomini] (1571?). *De Asia* [*Historia rerum ubique gestarum*]. In his *Opera quae extant omnia*, pp. 281–386. Basel: Henricpetrinus. Copy consulted: Trinity College Dublin, G.e.15.

PIUS II [Aeneas Silvius Piccolomini] (1984). *I commentarii*, ed. L. Totaro. Milano: Adelphi.

PIUS II [Aeneas Silvius Piccolomini] (2001). *De Europa*, ed. A. Van Heck. Città del Vaticano: Biblioteca Apostolica Vaticana.

PLINY [Gaius Plinius Secundus] (1942). *Natural History*, vol. 4 (Books 12–16), ed. & trans. H. Rackham. London – Cambridge, Mass.: Heinemann – Harvard University Press.

PONTANO, Giovanni (1902). *Urania*. In his *Carmina*, ed. B. Soldati, pp. 1–177. Firenze: Barbera.

PONTANO, Giovanni (1943a). *Dialogus qui Actius inscribitur*. In his *I dialoghi*, ed. C. Previtera, pp. 123–239. Firenze: Sansoni.

PONTANO, Giovanni (1943b). *Dialogus de ingratitudine qui Asinus inscribitur*. In his *I dialoghi*, ed. C. Previtera, pp. 289–308. Firenze: Sansoni.

PORCACCHI, Tommaso (1572). *L'isole più famose del mondo*. Venezia: Galignani & Porro. Copy consulted: Vatican Library, Cicognara VI. 2084 A.

PTOLEMY [Claudius Ptolemaeus] (1966). *Geographia*, ed. S. Münster, introd. R. A. Skelton. Amsterdam: Theatrum Orbis Terrarum. Anastatic reprint of Basel, Henricpetrinus, 1540 edition.

PTOLEMY [Claudius Ptolemaeus] (2000). J. L. Berggren and A. Jones (ed. & trans.), *Ptolemy's Geography. An Annotated Translation of the Theoretical Chapters*. Princeton – Oxford: Princeton University Press.

SACCHETTI, Franco (1996). 'Novella X'. In his *Il Trecentonovelle*, ed. V. Manucci, pp. 31–3. Roma: Salerno.

SILVESTRI, Domenico (1955). *De insulis et earum proprietatibus*, ed. C. Pecoraro. Palermo: *Atti dell'Accademia di Scienze, Lettere ed Arti di Palermo* 4 (14), 1953–4, 2 (2).

SOLINUS [Caius Iulius Solinus] (1512). *De situ orbis terrarum*. Pisaurum [Pesaro]: Soncino. Copy consulted: Roma, Biblioteca Nazionale, 6.21.Q.24.4.

SOLINUS [Caius Iulius Solinus] (1958). *Collectanea rerum memorabilium*, ed. T. Mommsen. Berlin: Weidmann.

STRABO (1988–97). *The Geography of Strabo*, vol. 1 (Books 1–2), vol. 2 (Books 3–5), vol. 6 (Books 13–14), ed. & trans. H. L. Jones. London – Cambridge, Mass.: Heinemann – Harvard University Press.

TACITUS [Cornelius Tacitus] (1938). *De origine et situ Germanorum*, ed. J. G. C. Anderson. Oxford: Clarendon.

TACITUS [Cornelius Tacitus] (1967). *De vita Agricolae*, ed. R. M. Ogilvie and I. Richmond. Oxford: Clarendon.

TASSO, Torquato (1967). *Gerusalemme liberata*, 2 vols, ed. L. Caretti. Bari: Laterza.

UBERTI, Fazio degli (1952). *Il dittamondo e le rime*, 2 vols, ed. G. Corsi. Bari: Laterza.

VARAGINE, Iacopo da [Jacobus a Voragine] (1969). *Legenda aurea*, ed. T. Graesse. Osnabrück: Otto Zeller.

VARAGINE, Iacopo da [Jacobus a Voragine] (1993). *The Golden Legend*, 2 vols., trans. W. G. Ryan. Princeton: Princeton University Press.

VIRGIL [Publius Vergilius Maro] (1998). *Eclogues, Georgics, Aeneid I – VI*, ed. & trans. H. R. Fairclough. London – Cambridge, Mass.: Heinemann – Harvard University Press.

VIRGILI, Polidoro [Polydore Vergil] (1534). *Anglicae historiae libri XXVI*. Basel: Bebelius. Copy consulted: Trinity College Dublin, P.b.5.

VIRGILI, Polidoro [Polydore Vergil] (2002). *On Discovery* [*De inventoribus rerum*], ed. & trans. B. P. Copenhaver. Cambridge, Mass. – London: Harvard University Press.

VIRGILI, Polidoro [Polydore Vergil] (2009). Description of Ireland from *Anglica historia*. Reproduced in Haywood (2009b), pp. 219–37.

Part Two: Secondary Texts

(No bibliography is ever complete. I had thought of limiting this section to works in English, but that would have given an unbalanced view of the state of knowledge in the field. I now include all the works to which I make specific reference in the text, as well as works I consider to be of more general relevance. I list the editions I consulted, in the language I consulted them, even though these may not always be the original, or first editions.)

Acta Sanctorum (1735). Martii, tomus II, ed. J. Bollandus et al. Venezia: Albrizzi & Coleti.

Agosti, B. (2008). *Paolo Giovio. Uno storico lombardo nella cultura artistica del Cinquecento*. Firenze: Olshki.

Alessio, G. C. (1990). 'The *lectura* of the *Triumphi* in the Fifteenth Century'. In K. Eisenbichler (ed.), *Petrarch's Triumphs. Allegory and Spectacle*, pp. 269–90. Ottawa: Center for Reformation and Renaissance Studies.

Allaire, G. (1993). 'The Secular Pilgrimage of an Errant Knight: Andrea da Barberino's *Guerrino il Meschino*', *Romance Languages Annual* 5, pp. 148–52.

Allaire, G. (1995). 'Tullia d'Aragona's *Il Meschino altramente detto il Guerrino* as Key to a Reappraisal of Her Work', *Quaderni d'Italianistica* 16, pp. 33–50.

Allaire, G. (1996). 'Portrayal of Muslims in Andrea da Barberino's *Guerrino il Meschino*'. In J. V. Tolan (ed.), *Medieval Christian Perceptions of Islam. A Book of Essays*, pp. 243–69. New York – London: Routledge.

Allaire, G. (1997). *Andrea da Barberino and the Language of Chivalry*. Gainesville, Florida: University Press of Florida.

Allaire, G. (2006). 'Narrating/Narrative Frontier Crossings in the Romance Epic *Guerrino Meschino*'. In O. Merisalo (ed.), *Frontiers in the Middle Ages*, pp. 629–36. Turnhout: Brepols.

Almagià, R. (1944). *Planisferi, carte nautiche e affini, dal secolo XIV al XVII, esistenti nella Biblioteca Apostolica Vaticana*. Città del Vaticano: Biblioteca Apostolica Vaticana.

Almagià, R. (1961). 'Osservazioni sull'opera geografica di Francesco Berlinghieri'. In his *Scritti geografici (1905–1957)*, pp. 497–526. Roma: Cremonese.

Anon. (1884). 'The *Dittamondo*. By a Discursive Contributor', *The Irish Monthly* 12, pp. 74–85 and 127–41.

Armstrong, L. (1996). 'Benedetto Bordon, *miniator*, and Cartography in Early Sixteenth-Century Venice', *Imago Mundi* 48, pp. 65–92.

Aujac, G. (1975). *La géographie dans le monde antique*. Paris: Presses Universitaires de France.

Aujac, G. (1993). 'La géographie grecque durant le Quattrocento: l'exemple de Strabon', *Geographia Antiqua* 2, pp. 147–69.

Babinger, F. (1963). 'Lorenzo de' Medici e la corte ottomana', *Archivio Storico Italiano* 121, pp. 305–61.

Bacchielli, R. (ed.) (2003). *Polidoro Virgili e la cultura umanistica europea*. Urbino: Accademia Raffaello.

Baldacci, O. (1992). 'La cultura geografica nel medioevo'. In *Optima hereditas. Sapienza giuridica romana e conoscenza dell'ecumene*, pp. 485–524. Milano: Scheiwiller.

Baldi, B. (2006). *Pio II e le trasformazioni dell'Europa cristiana (1457–1464)*. Milano: Edizioni Unicopli.

Baron, H. (1966). *The Crisis of the Early Italian Renaissance. Civic Humanism and Republican Liberty in an Age of Classicism and Tyranny*. Princeton: Princeton University Press.

Bartlett, R. (1982). *Gerald of Wales*. Oxford: Clarendon.

Bartoli, R. A. (1993). *La Navigatio Sancti Brendani e la sua fortuna nella cultura romanza dell'età del mezzo*. Fasano: Schena.

Basile, B. (1984). 'Sogni di terre lontane'. Chapter 6 of his *Poëta melancholicus*, pp. 325–68. Pisa: Pacini.

Berardis, V. (1950). *Italy and Ireland in the Middle Ages*. Dublin: Clonmore and Reynolds.

Bhabha, H. K. (2007). *The Location of Culture*. London – New York: Routledge.

Bieler, L. (1963). *Ireland, Harbinger of the Middle Ages*. London – New York – Toronto: Oxford University Press.

Billanovich, G. (1996). *Petrarca e il primo umanesimo*. Padova: Antenore.

Billanovich, M. (1968). 'Benedetto Bordon e Giulio Cesare Scaligero', *Italia Medievale e Umanistica* 9, pp. 187–256.

Boscolo, C. (1996–7). *L'immaginario della navigazione nel medioevo: il volgarizzamento veneto della Navigatio Sancti Brendani*. Università degli Studi di Trento, Facoltà di Lettere e Filosofia: unpublished *tesi di laurea*.

Bouloux, N. (2002). *Culture et savoirs géographiques en Italie au XIVe siècle*. Turnhout: Brepols.

Bradshaw, B., A. Hadfield and W. Maley (eds) (1993). *Representing Ireland. Literature and the Origins of Conflict, 1534–1660*. Cambridge: Cambridge University Press.

Branca, V. (1976). *Boccaccio. The Man and his Works*. New York: New York University Press.

Branca, V. (1994). 'Dal favoloso al realistico: esotismo fra pellegrini, mercanti e Boccaccio lanciati da pionieri sulle rotte di Colombo'. In R. Zorzi (ed.), *L'epopea delle scoperte*, pp. 39–59. Firenze: Olschki.

Brand, C. P. (1965). *Torquato Tasso. A Study of the Poet and of His Contribution to English Literature*. London: Cambridge University Press.

Brand, C. P. (1974). *Ludovico Ariosto. A Preface to the Orlando furioso*. Edinburgh: Edinburgh University Press.

Broc, N. (1996). *La geografia del Rinascimento. Cartografi, cosmografi, viaggiatori 1460–1520*, ed. C. Greppi. Modena: Panini.

Brotton, J. (1997). *Trading Territories. Mapping the Early Modern World*. London: Reaktion Books.

Brown, A. (1999). *The Renaissance*. London – New York: Longman.

Brown, A. (2000). 'The Language of Empire'. In W. J. Connell and A. Zorzi (eds), *Florentine Tuscany. Structures and Practices of Power*, pp. 32–47. Cambridge: Cambridge University Press.

Burckhardt, J. (1960). *The Civilization of the Renaissance in Italy*. New York: Mentor.

Burmeister, K.-H. (1964). *Sebastian Münster: eine Biographie*. Wiesbaden: Pressler.

Campbell, T. (1987). *The Earliest Printed Maps*. London: British Library.

Cardini, F. (1983). 'Orrizonti geografici e orizzonti mitici nel *Guerrin Meschino*'. In *Imago mundi. La conoscenza scientifica nel pensiero bassomedievale*, pp. 185–221. Todi: Accademia Tudertina.

Cardini, F. (1987). 'Mito del nord e conoscenza del settentrione europeo in alcune fonti fiorentine del Trecento'. Chapter 9 of his *Minima mediaevalia*, pp. 211–34. Firenze: Arnaud.

Cardini, F. (1990). 'Malatesta Ungaro al Purgatorio di San Patrizio'. In *Le signorie dei Malatesta. Atti della giornata di studi malatestiani a Sestino*, vol. 3, pp. 71–85. Rimini: Ghigi.

Carus-Wilson, E. M. (1967). *Medieval Merchant Venturers*. London: Methuen.

Casadei, A. (1988). *La strategia delle varianti: le correzioni storiche del terzo Furioso*. Lucca: Pacini Fazzi.

Cattaneo, A. (2011). *Fra Mauro's Mappa Mundi and Fifteenth-Century Venice*. Turnhout: Brepols.

Certeau, M. de (1986). *Heterologies. Discourse on the Other*, trans. B. Massumi, foreword W. Godzich. Manchester: Manchester University Press.

Cespedes, F. V. (1979). 'The Final Book of Polydore Vergil's *Anglica Historia*', *Viator* 10, pp. 375–96.

Chabod, F. (1971). *Storia dell'idea dell'Europa*. Bari: Laterza.

Cloulas, I. (1987). *Les Borgia*. Paris: Fayard.

Crane, N. (2003). *Mercator. The Man who Mapped the Planet*. London: Phoenix.

De Anna, L. (1994). *Il mito del nord. Tradizioni classiche e medievali*. Napoli: Liguori.

Defilippis, D. (1993). *Tradizione umanistica e cultura nobiliare nell'opera di Belisario Acquaviva*. Galatina: Congedo.

Defilippis, D. (ed.) (2009). *Da Flavio Biondo a Leandro Alberti. Corografia e antiquaria tra Quattro e Cinquecento*. Bari: Adriatica Editrice.

Delehaye, H. (1908). 'Le pélerinage de Laurent de Pászsthó au Purgatoire de S. Patrice', *Analecta Bollandiana* 27, pp. 35–60.

De Roo, P. (1924). *Material for a History of Pope Alexander VI, His Relatives and His Times*. Bruges: De Brouwer.

Descendre, R. (2009). *L'état du monde: Giovanni Botero entre raison d'état et géopolitique*. Genève: Droz.

Dictionnaire de théologie catholique vol. 13 (1936). Paris: Letouzey.

Diller, A. (1975). *The Textual Tradition of Strabo's Geography*. Amsterdam: Hakkert.

Edson, E. (2007). *The World Map, 1300–1492. The Persistence of Tradition and Transformation*. Baltimore: Johns Hopkins University Press.

Ellis, S. E. (1985). *Tudor Ireland. Crown, Community and the Conflict of Cultures 1470–1603*. London: Longman.

Fiorani, F. (2005). *The Marvel of Maps. Art, Cartography and Politics in Renaissance Italy*. New Haven – London: Yale University Press.

Firpo, L. (1971). 'Botero, Giovanni'. In *Dizionario biografico degli italiani*, vol. 13, pp. 352–62. Roma: Enciclopedia Treccani.

Fitzgibbon, A. (1876–8). 'Appendix to the Unpublished Geraldine Documents. The Gherardini of Tuscany', *Journal of the Royal Society of Antiquaries in Ireland* 14, pp. 246–64.

Foa, A. (1980). 'Chiericati, Francesco'. In *Dizionario biografico degli italiani*, vol. 24, pp. 674–81. Roma: Enciclopedia Treccani.

Frati, L. (1886). 'Il Purgatorio di San Patrizio secondo Stefano di Bourbon e Uberto da Romans', *Giornale Storico della Letteratura Italiana* 8, pp. 140–79.

Freeman, P. (2001). *Ireland and the Classical World.* Austin, Texas: University of Texas Press.
Fubini, R. (2006). 'Humanism and Scholasticism: Toward an Historical Definition'. In Mazzocco (2006), pp. 127–36.
Fueter, E. (1911). *Geschichte der neueren Historiographie.* München – Berlin: Oldenburg.
Gamurrini, E. (1671). *Istoria genealogica delle famiglie nobili toscane ed umbre*, vol. 2. Firenze: Guccio Navesi.
Garin, E. (1965). *Italian Humanism. Philosophy and Civic Life in the Renaissance.* Oxford: Blackwell.
Gautier Dalché, P. (2009). *La Géographie de Ptolémée en Occident (IVe–XVIe siècle).* Turnhout: Brepols.
Gentile, S. (ed.) (2000). *Sandro Botticelli pittore della 'Divina commedia'*, vol. 1. Milano. Exhibition catalogue.
Gentile, S. (2003). 'Umanesimo e cartografia: Tolomeo nel secolo XV'. In D. Ramada Curto, A. Cattaneo and A. Ferrand Almeida (eds), *La cartografia europea tra primo Rinascimento e fine dell'Illuminismo*, pp. 3–18. Firenze: Olschki.
Godi, C. (1996). *Bandello. Narratori e dedicatari della prima parte delle Novelle.* Roma: Bulzoni.
Gori, M. (1990). 'La geografia dell'epica tassiana', *Giornale Storico della Letteratura Italiana* 167, pp. 161–204.
Grafton, A., A. Shelford and N. Siraisi (1992). *New Worlds, Ancient Texts. The Power of Tradition and the Shock of Discovery.* Cambridge, Mass.: Harvard University Press.
Grignani, M. A. (ed.) (1975). *Navigatio Sancti Brendani. La navigazione di San Brendano.* Milano: Bompiani.
Guglielminetti, M. (1994). 'Il "Mondo Nuovo" come un'isola. Benedetto Bordon e Tommaso Porcacchi'. In R. Zorzi (ed.), *L'epopea delle scoperte*, pp. 103–20. Firenze: Olschki.
Hadfield, A. (1997). *Edmund Spenser's Irish Experience. Wilde Fruit and Salvage Soyl.* Oxford: Clarendon.
Hadfield, A. (ed.) (2004). *Amazons, Savages & Machiavels. Travel and Colonial Writing in English, 1550–1630: An Anthology.* Oxford: Oxford University Press.
Hadfield, A. and J. McVeagh (eds) (1994). *Strangers to That Land. British Perceptions of Ireland from the Reformation to the Famine.* Gerrards Cross: Colin Smythe.
Hainsworth, P. and D. Robey (eds) (2002). *The Oxford Companion to Italian Literature.* Oxford: Oxford University Press.
Hankey, A. T. (1957). 'Domenico de Bandino of Arezzo', *Italian Studies* 12, pp. 110–28.
Hankey, A. T. (1960). 'The Successive Revisions and Surviving Codices of the *Fons memorabilium universi* of Domenico di Bandino', *Rinascimento* 11, pp. 110–28.
Hankins, J. (ed.) (2003). *Renaissance Civic Humanism. Reappraisals and Reflections.* Cambridge: Cambridge University Press.

Hantzsch, V. (1965). *Sebastian Münster: Leben, Werk, wissenschaftliche Bedeutung.* Nieuwkoop: de Graaf.

Haren, M. and Y. de Pontfarcy (eds) (1988). *The Medieval Pilgrimage to St Patrick's Purgatory*. Enniskillen: Clogher Historical Society.

Harris, J. (2007). 'Ireland in Europe: Paolo Giovio's *Descriptio* (1548)', *Irish Historical Studies* 35, pp. 265–88.

Hay, D. (1952). *Polydore Vergil. Renaissance Historian and Man of Letters*. Oxford: Oxford University Press.

Hay, D. (1957). *Europe. The Emergence of an Idea*. Edinburgh: Edinburgh University Press.

Hay, D. and J. Law (1989). *Italy in the Age of the Renaissance, 1380–1530*. London – New York: Longman.

Hayes-McCoy, G. A. (1976). 'The Royal Supremacy and Ecclesiastical Revolution, 1534–47'. Chapter 2, vol. 3 (*Early Modern Ireland*) of *A New History of Ireland*, ed. T. W. Moody, F. X. Martin and F. J. Byrne, pp. 39–68. Oxford: Clarendon.

Haywood, E. (1997). '*Inter urinas liber factus est*. Il commento dell'Ilicino ai *Trionfi* del Petrarca.' In L. Rotondi Secchi Tarugi (ed.), *Petrarca e la cultura europea*, pp. 139–59. Milano: Nuovi Orizzonti.

Haywood, E. (1998). 'Paolo Giovio's *Descriptio Hyberniae*: Humanist Chorography or Political Manifesto?'. In J. F. Alcina et al. (ed.), *Acta conventus neo-latini bariensis*, pp. 315–22. Tempe, Arizona: Medieval & Renaissance Texts & Studies.

Haywood, E. (1999). '*La divisa dal mondo ultima Irlanda*, ossia la riscoperta umanistica dell'Irlanda', *Giornale Storico della Letteratura Italiana* 176, pp. 363–87.

Haywood, E. (2002). 'Disdegno umanista? Alessandro VI di fronte all'Irlanda'. In D. Canfora, M. Chiabò and M. de Nichilo (eds), *Principato ecclesiastico e riuso dei classici. Gli umanisti e Alessandro VI*, pp. 255–74. Roma: Roma nel Rinascimento.

Haywood, E. (2003a). 'Ariosto on Dante: Too Divine and Florentine'. In E. Haywood (ed.), *Dante Metamorphoses. Episodes in a Literary Afterlife*, pp. 71–104. Dublin: UCD Foundation for Italian Studies – Four Courts Press.

Haywood, E. (2003b). '*Iter asinarium*: per una nuova interpretazione dell'*Asinus* pontaniano'. In M. de Nichilo, G. Distaso and A. Iurilli (eds), *Confini dell'umanesimo letterario. Studi in onore di Francesco Tateo*, pp. 733–46. Roma: Roma nel Rinascimento.

Haywood, E. (2004). 'Letteratura militante? Il *Comento sopra la comedia* di Cristoforo Landino e la congiura dei Pazzi'. In L. Secchi Tarugi (ed.), *L'Europa del libro nell'età dell'Umanesimo*, pp. 93–111. Firenze: Franco Cesati.

Haywood, E. (2006). 'Il Petrarca lettore della *Topographia hibernica* di Giraldus Cambrensis'. In L. Secchi Tarugi (ed.), *Francesco Petrarca. L'opera latina: tradizione e fortuna*, pp. 647–67. Firenze: Franco Cesati.

Haywood, E. (2007). 'L'Europa senza isole di Enea Silvio Piccolomini'. In Secchi Tarugi (2007), pp. 237–60.

Haywood, E. (2009a). 'Geografia cristiana e geografia musulmana nella Firenze di Lorenzo il Magnifico'. In L. Secchi Tarugi (ed.), *Oriente e occidente nel Rinascimento*, pp. 93–106. Firenze: Franco Cesati.

Haywood, E. (2009b). 'Humanism's Priorities and Empire's Prerogatives: Polydore Vergil's Description of Ireland', *Proceedings of the Royal Irish Academy* 109C, pp. 195–237 (available online at: http://www.ria.ie/Publications/Journals/PRIA-Section-C/Online-access/109C-(2009).aspx).

Haywood, E. (2009c). '*Le rase carte degli indorati libri*: bibliofobia e bibliofilia nell'*Arcadia* di Iacopo Sannazaro'. In C. Corfiati and M. de Nichilo (eds), *Biblioteche nel Regno fra Tre e Cinquecento*, pp. 281–302. Lecce: Pensa MultiMedia.

Headley, J. M. (2000). 'Geography and Empire in the Late Renaissane: Botero's Assignment, Western Universalism, and the Civilising Process', *Renaissance Quarterly* 53, pp. 1119–55.

Hörnqvist, M. (2004). *Machiavelli and Empire*. Cambridge: Cambridge University Press.

Humboldt, A. von (1992). *L'invenzione del nuovo mondo: critica della conoscenza geografica*, ed. C. Greppi. Firenze: La Nuova Italia.

Jones, W. R. (1971). 'The Image of the Barbarian in Medieval Europe', *Comparative Studies in Society and History* 13, pp. 376–407.

Karrow, R. W. et al. (1998). *Abraham Ortelius (1527–1598), cartographe et humaniste*. Turnhout: Brepols.

Kelly, J. (2006). 'Lord Charlemont and Learning', *Proceedings of the Royal Irish Academy* 106C, pp. 395–407.

Kidwell, C. (1993). *Sannazaro and Arcadia*. London: Duckworth.

Kildare, Marquis of (1872). *The Earls of Kildare and Their Ancestors from 1057 to 1805*. Dublin: Hodges Smith & Co.

Kirkham, V. and A. Maggi (eds) (2009). *Petrarch. A Critical Guide to the Complete Works*. Chicago – London: University of Chicago Press.

Kraye, J. (ed.) (1996). *The Cambridge Companion to Renaissance Humanism*. Cambridge: Cambridge University Press.

Kristeller, P. O. (1980). 'Learned Women of Early Modern Italy: Humanists and University Scholars'. In P. H. Labalme (ed.), *Beyond Their Sex. Learned Women of the European Past*, pp. 91–116. New York: New York University Press.

Lastraioli, C. (2006). 'Un monde en forme d'île: espace géographique et espace imaginaire dans l'*Isolario* de Tommaso Porcacchi'. In A. Godard and M.-F. Piéjus (eds), *Espaces, histoire et imaginaire dans la culture italienne de la Renaissance*, pp. 43–76. Paris: Université de Paris III Sorbonne Nouvelle.

Le Goff, J. (1984). *The Birth of Purgatory*. London: Scolar.

Lestringant, F. (1991). 'Le déclin d'un savoir. La crise de la cosmographie à la fin de la Renaissance', *Annales ESC*, pp. 239–60.

Levy, F. J. (1967). *Tudor Historical Thought*. San Marino, Calif.: The Huntington Library.

Littara, V. (1997). *Storia di Noto antica (De rebus netinis)*, ed. & trans. F. Balsamo. Noto: Istituto per lo Studio e la Valorizzazione di Noto Antica.

Löwe, H. (ed.) (1982). *Die Iren und Europa im früheren Mittelalter*, 2 vols. Stuttgart: Klett-Cotta.

Mallett, M. E. (1969). *The Borgias. The Rise and Fall of a Renaissance Dynasty*. London: Bodley Head.

Martels, Z. von and A. Vanderjagt (eds) (2003). *Pius II: 'el più expeditivo pontifice'. Selected Studies on Aeneas Silvius Piccolomini (1405–1464)*. Leiden – Boston: Brill.

Martines, L. (2003). *April Blood. Florence and the Plot against the Medici*. London: Jonathan Cape.

Mayer, T. F. (1985). 'Reginald Pole in Paolo Giovio's *Descriptio*: A Strategy for Reconversion', *The Sixteenth Century Journal* 16, pp. 431–50.

Mazzocco, A. (ed.) (2006). *Interpretations of Renaissance Humanism*. Leiden – Boston: Brill.

McGuire, B. P. (1989). 'Purgatory, the Communion of Saints and Medieval Change', *Viator* 20, pp. 61–84.

McLaughlin M., L. Panizza and P. Hainsworth (eds) (2007). *Petrarch in Britain. Interpreters, Imitators, and Translators over 700 Years*. Oxford: Oxford University Press (Proceedings of the British Academy 146).

McLean, M. (2007). *The Cosmographia of Sebastian Münster: Describing the World in the Reformation*. Aldershot: Ashgate.

Meehan, C. P. (1878). *The Rise, Increase and Exit of the Geraldines, Earls of Desmond*. Dublin: J. Duffy.

Melis, F. (1990). *I mercanti italiani dell'Europa medievale e rinascimentale*. Firenze: Le Monnier.

Menetti, E. (2000). 'Le inquietudini di un narratore: Matteo Bandello'. In G. Albanese (ed.), *Favole, parabole, istorie. Le forme della scrittura novellistica dal medioevo al Rinascimento*, pp. 439–64. Roma: Salerno.

Merry, V. (1986). 'Una nota sulla fortuna del commento di Bernardo Ilicino ai *Trionfi* petrarcheschi', *Giornale Storico della Letteratura Italiana* 163, pp. 235–46.

Michelacci, L. (2004). *Giovio in Parnaso: tra collezione di forme e storia universale*. Bologna: Il Mulino.

Milanesi, M. (1992). 'La rinascita della geografia dell'Europa (1350–1480)'. In S. Gensini (ed.), *Europa e Mediterraneo tra medio evo e prima età moderna. L'osservatorio italiano*, pp. 35–59. Pisa: Pacini.

Milanesi, M. (1993). 'Il *De insulis et earum proprietatibus* di Domenico Silvestri (1385–1406)', *Geographia Antiqua* 2, pp. 133–46.

Milanesi, M. (1994). 'Il commento al *Dittamondo* di Guglielmo Capello (1435–1437)'. In M. Bertozzi (ed.), *Alla corte degli Estensi. Filosofia, arte e cultura a Ferrara nei secoli XV e XVI*, pp. 365–88. Ferrara: Università degli Studi.

Mitchell, R. J. (1962). *The Laurels and the Tiara. Pope Pius II, 1458–1464*. London: Harvill.

Moody, T. W., F. X. Martin and F. J. Byrne (eds) (1976). *A New History of Ireland* vol. 3. Oxford: Oxford University Press.

Morgan, H. (1999). 'Giraldus Cambrensis and the Tudor Conquest of Ireland'. In H. Morgan (ed.), *Political Ideology in Ireland, 1541–1641*, pp. 22–44. Dublin: Four Courts Press.

Morsolin, B. (1873). *Francesco Chiericati, vescovo di Vicenza*. Vicenza: Accademia Olimpica.

Muldoon, J. (2003). 'Wild Irishmen and Half-Breeds'. In his *Identity on the Medieval Irish Frontier. Degenerate Englishmen, Wild Irishmen, Middle Nations*, pp. 63–102. Gainesville, Florida: University Press of Florida.

Mund-Dopchie, M. (2009). *Ultima Thulé: histoire d'un lieu et genèse d'un mythe*. Genève: Droz.

Najemy, J. M. (ed.) (2004). *Italy in the Age of the Renaissance, 1300–1500*. Oxford: Oxford University Press.

New Catholic Encyclopaedia vol. 11 (1967). New York: McGraw-Hill.

O'Brien, E. (2009). 'Arms and Letters: Julius Caesar, the *Commentaries* of Pope Pius II, and the Politicization of Papal Imagery', *Renaissance Quarterly* 62, pp. 1057–97.

Osella, G. (1932). *Il Guerrin Meschino*. Torino: Giovanni Chiantore.

O'Sullivan, M. D. (1962). *Italian Merchant Bankers in Ireland in the Thirteenth Century*. Dublin: Figgis.

Pagden, A. (1993). *European Encounters with the New World. From Renaissance to Romanticism*. New Haven – London: Yale University Press.

Pagden, A. (2002). 'Europe: Conceptualizing a Continent'. In A. Pagden (ed.), *The Idea of Europe from Antiquity to the European Union*, pp. 33–54. Cambridge: Cambridge University Press.

Palazzetti, C. (1998). *Il pozzo della rocca di San Patrizio, Orvieto*. Viterbo: BetaGamma.

Pastore Stocchi, M. (1963). *Tradizione medievale e gusto umanistico nel 'De montibus' del Boccaccio*. Padova, CEDAM.

Pastore Stocchi, M. (1992). 'La cultura geografica dell'Umanesimo'. In *Optima hereditas. Sapienza giuridica romana e conoscenza dell'ecumene*, pp. 563–86. Milano: Scheiwiller.

Pellegrini, M. (2000). 'Pio II', in *Enciclopedia dei papi*. http://www.treccani.it/enciclopedia/pio-ii_%28Enciclopedia-dei-Papi%29/ (accessed 16 June 2013).

Picard, J.-M. (1988). 'The Italian Pilgrims'. Chapter 7 of Haren and Pontfarcy (1988), pp. 169–89.

Pratt, M. L. (1992). *Imperial Eyes. Travel Writing and Transculturation*. London – New York: Routledge.

Prontera, F. (1992). 'La cultura geografica in età imperiale'. In *Optima hereditas. Sapienza giuridica romana e conoscenza dell'ecumene*, pp. 277–317. Milano: Scheiwiller.

Quondam, A. (1985). '(De)scrivere la terra. Il discorso geografico da Tolomeo all'Atlante'. In *Culture et société en Italie du moyen-âge à la Renaissance. Hommage a André Rochon*, pp. 11–35. Paris: Université de la Sorbonne Nouvelle.

Ragni, E. (1986). 'Fazio degli Uberti'. In V. Branca (ed.), *Dizionario critico della letteratura italiana*, 2nd ed., vol. 1, pp. 224–9. Torino: UTET.

Randles, W. G. L. (1994). 'Classical Models of World Geography and Their Transformation following the Discovery of America'. In W. Haase and M. Reinhold (eds), *The Classical Tradition and the Americas*, vol. 1: *European Images of the Americas and the Classical Tradition*, part 1, pp. 5–76. Berlin – New York: de Gruyter.

Ricci, P. G. (1950). 'Per una monografia su Domenico Silvestri', *Annali della Scuola Normale Superiore di Pisa. Lettere, Storia e Filosofia* 2 (19), pp. 13–24.

Rollo, D. (1995). 'Gerald of Wales' *Topographia Hibernica*: Sex and the Irish Nation', *Romanic Review* 86, pp. 169–90.

Rossi, A. (1984). 'Filologia, grammatica e retorica negli scrittoi trecenteschi', *Poliorama* 3, pp. 3–22.

Rubiés, J.-P. (2000). *Travel and Ethnology in the Renaissance. South India through European Eyes, 1250–1625*. Cambridge: Cambridge University Press.

Rubiés, J.-P. (2007). *Travellers and Cosmographers. Studies in the History of Early Modern Travel and Ethnology*. Aldershot: Ashgate.

Said, E. (1993). *Culture and Imperialism*. New York: Knopf.

Salmeri, G. (1988). 'Tra politica e antiquaria: lettura di Strabone nel XV e XVI secolo'. In G. Maddoli (ed.), *Strabone e l'Italia antica*, pp. 289–312. Napoli: Edizioni Scientifiche Italiane.

Salvadori, C. (2003). 'Landmarks in the Fortunes of Dante in the Florentine Quattrocento'. In E. Haywood (ed.), *Dante Metamorphoses. Episodes in a Literary Afterlife*, pp. 43–70. Dublin: UCD Foundation for Italian Studies – Four Courts Press.

Scherman, K. (1981). *The Flowering of Ireland. Saints, Scholars and Kings*. London: Gollancz.

Secchi Tarugi, L. (ed.) (2007). *Pio II umanista europeo*. Firenze: Franco Cesati.

Seymour, St J. D. (1918). *St Patrick's Purgatory. A Medieval Pilgrimage in Ireland*. Dundalk: Tempest.

Shirley, J. W. and F. D. Hoeniger (1985). *Science and the Arts in the Renaissance*. Washington – London – Toronto: Associated University Presses.

Sigismondo de' Conti da Foligno (1883). *Le storie de' suoi tempi dal 1475 al 1510.* Roma: Barbera.

Smyth, B. (ed.) (1966). *The Register of Milo Sweetman, Archbishop of Armagh.* Dublin: Irish Manuscripts Commission.

Stoppa, A. L. and R. Cicala (eds) (1992). *L'umanista aronese Pietro Martire d'Anghiera, primo storico del nuovo mondo.* Novara: Associazione di Storia della Chiesa Novarese.

Tateo, F. (2002). 'Un'orazione umanistica per la storia antica di Taranto (Il *De antiquitate et varia fortuna Tarentinorum* di Giovanni Giovane)', *La Nuova Ricerca* 11, pp. 181–93.

Tateo, F. and D. Canfora (eds) (2004). *Lorenzo, Poliziano, Sannazaro (Poggio e Pontano).* Roma: Istituto Poligrafico e Zecca dello Stato.

Taylor, J. (1966). *The 'Universal Chronicle' of Ranulf Higden.* Oxford: Clarendon.

Tierney, J. J. (1959–60). 'The Celtic Ethnography of Posidonius', *Proceedings of the Royal Irish Academy* 60, pp. 189–275.

Tiraboschi, G. (1784). *Storia della letteratura italiana*, tome 7, part 2 (vol. 9). Roma: Perego Salviani.

Tommasini, A. M. (1937). *Irish Saints in Italy.* London – Glasgow: Sands & Co.

Tortoli, G. (ed.) (1904). *Rime pie edite e inedite di Messer Dolcibene.* Prato: Passerini.

Vickers, B. (1990). 'Leisure and Idleness in the Renaissance: The Ambivalence of *otium*', *Renaissance Studies* 4, pp. 1–37 and 107–54.

Veneziani, P. (1982). 'Vicende topografiche della *Geografia* di Francesco Berlinghieri', *La bibliofilia* 84, pp. 195–208.

Ware, J. (1764). *The Whole Works*, vol. 1, ed. W. Harris. Dublin: Bell & Fleming.

Wilkins, D. (1964). *Concilia Magnae Britanniae et Hiberniae* vol. 3. Bruxelles: Culture et Civilisation.

Wilkins, E. H. (1961). *The Life of Petrarch.* Chicago – London: Chicago University Press.

Woodhouse, J. (1978). *Baldesar Castiglione. A Reassessment of 'The Courtier'.* Edinburgh: Edinburgh University Press.

Zak, G. (2010). *Petrarch's Humanism and the Care of the Self.* Cambridge: Cambridge University Press.

Zimmermann, T. C. P. (1995). *Paolo Giovio. The Historian and the Crisis of Sixteenth-Century Italy.* Princeton: Princeton University Press.

Index

This Index is divided into six sections: Authors – Fictional and Mythical Characters – Historical Characters and Events – Ireland – Places – Themes. When cross-references are to another section, that section is specified. If there is no specification, the cross-reference is to the same section. Characters who could belong in two sections (e.g. Charlemagne) are only ever fully listed in one, with a cross-reference to the other. The names of fictional characters are followed, in brackets, by the name of the author(s) in whose work(s) they appear.

Authors

Fictional and Mythical Characters

Historical Characters and Events

Ireland

Places

Themes

www.ingramcontent.com/pod-product-compliance
Ingram Content Group UK Ltd.
Pitfield, Milton Keynes, MK11 3LW, UK
UKHW021852100926
8556IPUK00020B/1186

9 783034 317580